Praise for

MICHIO KAKU and
QUANTUM SUPREMACY

"Kaku is an international treasure." —*The Times* (London)

"[Kaku is] a masterful science communicator. . . . A consummate storyteller." —*The Wall Street Journal*

"Kaku . . . has the rare ability to take complicated scientific theories and turn them into readable tales about what our lives will be like in the future." —*USA Today*

"Mind-blowing. . . . Translating complicated scientific concepts into language that lay readers can understand is an art. Kaku . . . is one of the best practitioners. . . . As always, Kaku's enthusiasm is contagious, and this latest book is an important guide to a crucial part of the tech future. An informative and highly entertaining read about the computing revolution already underway." —*Kirkus Reviews* (starred review)

"Illuminating. . . . Revealing a breathtaking, expansive look into the promise, power, and possibility of quantum computing. . . . Kaku will spark the imagination of [readers] interested in the nexus of computers and quantum mechanics." —*Booklist*

Michio Kaku

QUANTUM SUPREMACY

Michio Kaku is a professor of theoretical physics at the City University of New York, cofounder of string field theory, and the author of several widely acclaimed science books, including *Hyperspace*, *Beyond Einstein*, *Physics of the Impossible*, *Physics of the Future*, *The Future of the Mind*, *The Future of Humanity*, and *The God Equation*. He was the science correspondent for CBS's *This Morning*, and he's the host of the radio programs *Science Fantastic* and *Exploration* as well as of several science TV specials for the BBC and the Discovery and Science Channels.

mkaku.org

QUANTUM SUPREMACY

How the Quantum Computer Revolution
Will Change Everything

DR. MICHIO KAKU

PROFESSOR OF THEORETICAL PHYSICS

CITY UNIVERSITY OF NEW YORK

VINTAGE BOOKS

A DIVISION OF PENGUIN RANDOM HOUSE LLC

NEW YORK

The Library of Congress has cataloged the Doubleday edition as follows:
Name: Kaku, Dr. Michio, author.
Title: Quantum supremacy : how the quantum computer
revolution will change everything / Dr. Michio Kaku, Professor
of Theoretical Physics, City University of New York.
Description: First edition. | New York : Doubleday, 2023. |
Includes bibliographical references and index.
Identifiers: LCCN 2022046826 (print) | LCCN 2022046827 (ebook)
Subjects: LCSH: Quantum computers. | Quantum computing.
Classification: LCC QA76.889 .K36 2023 (print) |
LCC QA76.889 (ebook) | DDC 006.3/843—dc23
LC record available at https://lccn.loc.gov/2022046826
LC ebook record available at https://lccn.loc.gov/2022046827

Vintage Books Trade Paperback ISBN: 978-0-593-46700-8
eBook ISBN: 978-0-385-54837-3

Book design by Michael Collica

vintagebooks.com

Printed in the United States of America
10   9   8   7   6   5   4   3   2   1

To my loving wife, Shizue, and my daughters,
Dr. Michelle Kaku and Alyson Kaku

CONTENTS

Part IV: Modeling the World and the Universe

PART I

RISE OF QUANTUM COMPUTERS

CHAPTER 1

END OF THE AGE OF SILICON

A revolution is coming.

In 2019 and 2020, two bombshells rocked the world of science. Two groups announced that they had achieved quantum supremacy, the fabled point at which a radically new type of computer, called a quantum computer, could decisively outperform an ordinary digital supercomputer on specific tasks. This heralded an upheaval that can change the entire computing landscape and overturn every aspect of our daily life.

First, Google revealed that their Sycamore quantum computer could solve a mathematical problem in 200 seconds that would take 10,000 years on the world's fastest supercomputer. According to MIT's *Technology Review*, Google called this a major breakthrough. They likened it to the launch of Sputnik or the Wright brothers' first flight. It was "the threshold of a new era of machines that would make today's mightiest computer look like an abacus."

Then the Quantum Innovation Institute at the Chinese Academy of Sciences went even further. They claimed their quantum computer was 100 trillion times faster than an ordinary supercomputer.

IBM vice president Bob Sutor, commenting on the meteoric rise of quantum computers, states flatly, "I think it's going to be the most important computing technology of this century."

Quantum computers have been called the "Ultimate Computer," a decisive leap in technology with profound implications for the entire world. Instead of computing on tiny transistors, they compute on the tiniest possible object, the atoms themselves, and hence can easily surpass the power of our greatest supercomputer. Quantum computers might usher in an entirely new age for the economy, society, and our way of life.

But quantum computers are more than just another powerful computer. They are a new type of computer that can tackle problems that digital computers can never solve, even with an infinite amount of time. For example, digital computers can never accurately calculate how atoms combine to create crucial chemical reactions, especially those that make life possible. Digital computers can only compute on digital tape, consisting of a series of 0s and 1s, which are too crude to describe the delicate waves of electrons dancing deep inside a molecule. For example, when tediously computing the paths taken by a mouse in a maze, a digital computer has to painfully analyze each possible path, one after the other. A quantum computer, however, *simultaneously* analyzes all possible paths at the same time, with lightning speed.

This in turn has heightened an intense rivalry between competing computer giants, which are all racing to create the world's most powerful quantum computer. In 2021, IBM unveiled its own quantum computer, called the Eagle, which has taken the lead, with more computing power than all previous models.

But these records are like pie crusts—they are made to be broken.

Given the profound implications of this revolution, it is not surprising that many of the world's leading corporations have invested heavily in this new technology. Google, Microsoft, Intel, IBM, Rigetti, and Honeywell are all building quantum computer prototypes. The leaders of Silicon Valley realize that they must keep pace with this revolution or be left in the dust.

IBM, Honeywell, and Rigetti Computing have put their first-generation quantum computers on the internet to whet the appetite of a curious public, so that people may gain their first direct exposure to quantum computation. One can experience this new quantum revolution firsthand by connecting to a quantum computer on the internet. For example, the "IBM Q Experience," launched in 2016, makes fifteen quantum computers available to the public via the internet for free. Samsung and JPMorgan Chase are among these users. Already, 2,000 people, from schoolchildren to professors, use them every month.

Wall Street has taken a keen interest in this technology. IonQ became the first major quantum computing company to go public, raising $600 million in its IPO in 2021. Even more startling, the rivalry is so intense that a new start-up, PsiQuantum, without any commercial prototype on the market or any track record of previous products, suddenly soared on Wall Street to a $3.1 billion valuation, with the ability to capture $665 million in funding almost overnight. Business analysts wrote that they had rarely seen anything like this, a new company riding the tide of feverish speculation and sensational headlines to such heights.

Deloitte, the consulting and accounting firm, estimates that the market for quantum computers should reach hundreds of millions of dollars in the 2020s and tens of billions of dollars in the 2030s. No one knows when quantum computers will enter the commercial marketplace and alter the economic landscape, but predictions are being revised all the time to match the unprecedented speed of scientific discovery in this field. Christopher Savoie, CEO of Zapata Computing, speaking about the meteoric rise of quantum computers, says, "It's no longer a matter of if, but when."

Even the U.S. Congress has expressed keen interest in helping jump-start this new quantum technology. Realizing that other nations have already generously funded research in quantum computers, in December 2018, Congress passed the National Quantum Initiative Act to provide seed money to help spark new research. It mandated the formation of two to five new National Quantum Information Science Research Centers, to be funded with $80 million annually.

In 2021, the U.S. government also announced an investment of $625 million in quantum technologies, to be supervised by the Department of Energy. Giant corporations like Microsoft, IBM, and Lockheed Martin also contributed an additional $340 million to this project.

China and the U.S. are not the only ones using government funds to accelerate this technology. The U.K. government is now constructing the National Quantum Computing Centre, which will serve as a hub for research on quantum computing, to be built at the Harwell lab of the Science and Technology Facilities Council in Oxfordshire. Spurred on by the government, there were thirty quantum computer start-ups founded in the U.K. by the end of 2019.

Industry analysts recognize that it's a trillion-dollar gamble. There are no guarantees in this highly competitive field. Despite the impressive technical achievements made by Google and others in recent years, a workable quantum computer that can solve real-world problems is still many years in the future. A mountain of hard work still lies before us. Some critics even claim it could be a wild-goose chase. But computer companies realize that unless they have a foot in the door, it might slam shut on them.

Ivan Ostojic, a partner at consulting firm McKinsey, says, "Companies in the industries where quantum will have the greatest potential for complete disruption should get involved in quantum right now." Areas like chemistry, medicine, oil and gas, transportation, logistics, banking, pharmaceuticals, and cybersecurity are ripe for major change. He adds, "In principle, quantum will be relevant for all CIOs as it will accelerate solutions to a large range of problems. Those companies need to become owners of quantum capability."

Vern Brownell, former CEO of D-Wave Systems, a Canadian quantum computing company, remarks, "We believe we're right on the cusp of providing capabilities you can't get with classical computing."

Many scientists believe that we are now entering an entirely new era, with shock waves comparable to those created by the introduction of the transistor and the microchip. Companies without direct ties to computer production, like the automotive giant Daimler, which owns Mercedes-Benz, are already investing in this new technology, sensing that quantum computers may pave the way for new developments in their own industries. Julius Marcea, an executive with rival BMW, has written, "We are excited to investigate the transformative potential of quantum computing on the automotive industry and are committed to extending the limits of engineering performance." Other large companies, like Volkswagen and Airbus, have set up quantum computing divisions of their own to explore how this may revolutionize their business.

Pharmaceutical companies are also watching developments in this field intently, realizing that quantum computers may be able to simulate complex chemical and biological processes that are far beyond the capability of digital computers. Huge facilities devoted to testing millions of drugs may one day be replaced by "virtual laboratories" that test drugs in cyberspace. Some have feared that perhaps this might one day replace chemists. But Derek Lowe, who runs a blog about drug discovery, says, "It is not that machines are going to replace chemists. It's that the chemists who use machines will replace those that don't."

Even the Large Hadron Collider outside Geneva, Switzerland, the biggest science machine in the world, which slams protons together at 14 trillion electron volts to re-create the conditions of the early universe, now uses quantum computers to help sift through mountains of data. In one second, they can analyze up to one trillion bytes generated by about one billion particle collisions. Perhaps one day quantum computers will unravel the secrets of the creation of the universe.

Quantum Supremacy

Back in 2012, when physicist John Preskill of the California Institute of Technology first coined the term "quantum supremacy," many scientists shook their heads. It would take decades, if not centuries, they thought, before quantum computers could outperform a digital computer. After all, computing on individual atoms, rather than wafers of silicon chips, was considered fiendishly difficult. The slightest vibration or noise can disturb the delicate dance of atoms in a quantum computer. But these stunning announcements of quantum supremacy have so far shredded naysayers' gloomy predictions. Now the concern is shifting to how fast the field is developing.

The tremors caused by these remarkable achievements have also shaken boardrooms and top secret intelligence agencies around the world. Documents leaked by whistleblowers have shown that the CIA and the National Security Agency are closely following developments in the field. This is because quantum computers are so powerful that, in principle, they could break all known cybercodes. This means that the secrets carefully guarded by governments, which are their crown jewels containing their most sensitive information, are vulnerable to attack, as are the best-kept secrets of corporations and even individuals. This situation is so urgent that even the U.S. National Institute of Standards and Technology (NIST), which sets national policy and standards, recently issued guidelines to help large corporations and agencies plan for the inevitable transition to this new era. NIST has already announced they expect that by 2029 quantum computers will be able to break 128-bit AES encryption, the code used by many companies.

Writing in *Forbes* magazine, Ali El Kaafarani notes, "That's a pretty terrifying prospect for any organization with sensitive information to protect."

The Chinese have spent $10 billion on their National Laboratory for Quantum Information Sciences because they are determined to be a leader in this vital, fast-moving field. Nations spend tens of billions to jealously guard these codes. Armed with a quantum computer, a hacker might conceivably break into *any* digital computer on the planet, thereby disrupting industries and even the military. All sensitive information may become available to the highest bidder. Financial markets might also be thrown into turmoil by quantum computers breaking into the inner sanctum of Wall Street. Quantum computers might also unlock the blockchain, creating havoc in the bitcoin market as well. Deloitte has estimated that about 25 percent of bitcoins are potentially vulnerable to hacking by a quantum computer.

"Those running blockchain projects will likely be keeping a nervous eye on quantum computing advancements," concludes a report by CB Insights, a data software IT company.

So what is at stake is nothing less than the world economy, which is heavily wedded to digital technology. Wall Street banks use computers to keep track of multibillion-dollar transactions. Engineers use computers to design skyscrapers, bridges, and rockets. Artists depend on computers to animate Hollywood blockbusters. Pharmaceutical companies use computers to develop their next wonder drug. Children rely on computers to play the latest video game with their friends. And we crucially depend on cell phones to give us instantaneous news from our friends, associates, and relatives. All of us have had the experience of being thrown into a panic when we cannot find our cell phone. In fact, it is extremely difficult to name any human activity that hasn't been turned upside down by computers. We are so dependent on them that if somehow all the world's computers suddenly came to an abrupt halt, civilization would be thrown into chaos. That is why scientists are following the development of quantum computers so intently.

End of Moore's Law

What is driving all this turmoil and controversy?

The rise of quantum computers is a sign that the Age of Silicon is gradually coming to a close. For the past half-century, the explosion of computer power has been described by Moore's law, named after Intel founder Gordon Moore. Moore's law states that computer power doubles every eighteen months. This deceptively simple law has tracked the remarkable exponential increase in computer power, which is unprecedented in human history. There is no other invention which has had such a pervasive impact in such a brief period of time.

Computers have gone through many stages throughout their history, each time vastly increasing their power and causing major societal change. Moore's law, in fact, can be extended all the way back to the 1800s, to the age of mechanical computers. Back then, engineers used spinning cylinders, cogs, gears, and wheels to perform simple arithmetic operations. At the turn of the last century, these calculators began to use electricity, replacing gears with relays and cables. During World War II, computers used vast arrays of vacuum tubes to break secret government codes. In the postwar era, the transition was made from vacuum tubes to transistors, which could be miniaturized to microscopic size, facilitating continued advances in speed and power.

Back in the 1950s, mainframe computers could only be purchased by large corporations and government agencies like the Pentagon and international banks. They were powerful (for example, the ENIAC could do in thirty seconds what might take a human twenty hours). But they were expensive, bulky, and often took up an entire floor of an office building. The microchip revolutionized this entire process, decreasing in size over the decades so that a typical chip the size of your fingernail can now contain about one billion transistors. Today, cell phones used by children to play video games are more powerful than a roomful of those lumbering dinosaurs once used by the Pentagon. We take for granted that the computer in our pocket exceeds the power of the computers used during the Cold War.

All things must pass. Each transition in the development of the computer rendered the previous technology obsolete in a process of creative destruction. Moore's law is already slowing down and may eventually come to a halt. This is because microchips are so compact that the thinnest layer of transistors is about twenty atoms across. When they reach about five atoms across, the location of the electron becomes uncertain, and they can leak out and short-circuit the chip or generate so much heat that the chips melt. In other words, by the laws of physics, Moore's law must eventually collapse if we continue to use primarily silicon. We could be witnessing the end of the Age of Silicon. The next leap might be the post-Silicon or Quantum Age.

As Intel's Sanjay Natarajan has said, "We've squeezed, we believe, everything you can squeeze out of that architecture."

Silicon Valley may eventually become the next Rust Belt.

Although things seem calm now, sooner or later this new future will dawn. As Hartmut Neven, director of Google's AI lab, says, "It looks like nothing is happening, nothing is happening, and then whoops, suddenly you're in a different world."

Why Are They So Powerful?

What makes quantum computers so powerful that the nations of the world are rushing to master this new technology?

Essentially, all modern computers are based on digital information, which can be encoded in a series of 0s and 1s. The smallest unit of information, a single digit, is called a bit. This sequence of 0s and 1s is fed into a digital processor, which performs the calculation, and then produces an output. For example, your internet connection may be measured in terms of gigabits per second (Gbit/s) which means that one billion bits are being sent to your computer every second, giving you instant access to movies, emails, documents, etc.

However, Nobel laureate Richard Feynman in 1959 saw a different approach to digital information. In a prophetic, pathbreaking essay titled "There's Plenty of Room at the Bottom" and subsequent articles, he asked: Why not replace this sequence of 0s and 1s with states of atoms, making an atomic computer? Why not replace transistors with the smallest possible object, the atom?

Atoms are like spinning tops. In a magnetic field, they can align either up or down with respect to the magnetic field, which can correspond to a 0 or a 1. The power of a digital computer is related to the number of states (the 0s or 1s) you have in your computer.

But due to the weird rules of the subatomic world, atoms can also spin in any combination of the two. For example, you can have a state in which the atom spins up 10 percent of the time and spins down 90 percent of the time. Or it spins up 65 percent of the time and spins down 35 percent of the time. In fact, there are an infinite number of ways that you can have an atom spin. This vastly increases the number of states that are possible. So the atom can carry much more information, not just in a bit, but a qubit, i.e., a simultaneous mixture of the up and down states. Digital bits can only carry one bit of information at a time, which limits their power, but qubits, or quantum bits, have almost unlimited power. The fact that, at the atomic level, objects can exist simultaneously in multiple states is called superposition. (This also means the familiar laws of common sense are routinely violated at the atomic level. At that scale, electrons can be in two places at the same time, which is not true for large objects.)

In addition, these qubits can interact with each other, which is not possible for ordinary bits. This is called entanglement. Whereas digital bits have independent states, each time you add another qubit, it interacts with all the previous qubits, so you double the number of possible interactions. Hence, quantum computers are inherently exponentially more powerful than digital computers, because you double the number of interactions every time you add an additional qubit.

For example, today quantum computers can have over 100 qubits. This means that they are 2^{100} times more powerful than a supercomputer with just one qubit.

Google's Sycamore quantum computer, which was the first to achieve quantum supremacy, has the power to process 72 million billion bytes of memory with its fifty-three qubits. So a quantum computer like Sycamore dwarfs any conventional computer.

The commercial and scientific implications of this are enormous. As we transition from a digital world economy to a quantum economy, the stakes are extraordinarily high.

Speed Bumps to Quantum Computers

The next key question is: What prevents us from marketing powerful quantum computers today? Why doesn't some enterprising inventor unveil a quantum computer that can break any known code?

The problem facing quantum computers was also foreseen by Richard Feynman when he first proposed the concept. In order for quantum computers to work, atoms have to be arranged precisely so that they vibrate in unison. This is called coherence. But atoms are incredibly small and sensitive objects. The smallest impurity or disturbance from the outside world can cause this array of atoms to fall out of coherence, ruining the entire calculation. This fragility is the main problem facing quantum computers. So the trillion-dollar question is: Can we control decoherence?

In order to minimize the contamination coming from the outside world, scientists use special equipment to drop the temperature to near absolute zero, where unwanted vibrations are at a minimum. But this requires expensive, special pumps and tubing to reach those temperatures.

But we are faced with a mystery. Mother Nature uses quantum mechanics at room temperature without a problem. For example, the miracle of photosynthesis, one of the most important processes on earth, is a quantum process, yet it takes place at normal temperatures. Mother Nature does not use a roomful of exotic devices operating at near absolute zero to execute photosynthesis. For reasons that are not well understood, in the natural world coherence can be maintained even on a warm, sunny day, when disturbances from the outside world should create chaos at the atomic level. If we could one day figure out how Mother Nature performs her magic at room temperature, then we might become masters of the quantum and even life itself.

Revolutionizing the Economy

Although quantum computers pose a threat to the cybersecurity of nations in the short term, they also have vast practical implications in the long term, with the power to revolutionize the world economy, create a more sustainable future, and usher in an era of quantum medicine to help cure previously incurable diseases.

There are many areas where quantum computers can overtake conventional digital computers:

1. Search engines

In the past, wealth might be measured in terms of oil or gold.

Now, increasingly it is measured in data. Companies used to throw their own financial data away, but now this information is being recognized as more valuable than precious metals. But sifting through mountains of data may overwhelm a conventional digital computer. This is where quantum computers come in, by finding the needle in the haystack. Quantum computers may be able to analyze a company's finances in order to isolate the handful of factors that are preventing it from growing.

In fact, JPMorgan Chase has recently partnered with IBM and Honeywell in order to analyze its data to make better predictions of financial risk and uncertainty and increase the efficiency of their operations.

2. Optimization

Once quantum computers have used search engines to identify the key factors in the data, the next question is how to adjust them to maximize certain factors, such as profit. At the very least, large corporations, universities, and government agencies will use quantum computers to minimize their expenses and maximize their efficiency and profit. For example, a company's net proceeds depends on hundreds of factors, such as salaries, sales, expenses, and so forth, which all change rapidly with time. It might overwhelm a traditional digital computer to find the right combination of these myriad factors to maximize their profit margin. Meanwhile, a financial firm may want to use quantum computers to predict the future of certain financial markets that handle billions of dollars in transactions daily. This is where quantum computers can help by providing the computational muscle to optimize their bottom line.

3. Simulation

Quantum computers might also solve complex equations that are beyond the ability of digital computers. For example, engineering firms may use quantum computers to calculate the aerodynamics of jets, airplanes, and cars, to find the ideal shape that reduces friction, minimizes cost, and maximizes efficiency. Or governments may use quantum computers to predict the weather, from determining the path of a monster hurricane to calculating how global warming will affect the economy and our way of life decades into the future. Or scientists may use quantum computers to find the optimal configuration of magnets in giant nuclear fusion machines in order to harness the power of hydrogen fusion and "put the sun in a bottle."

But perhaps the greatest benefit is to use quantum computers to simulate hundreds of vital chemical processes. The dream would be to predict the outcome of any chemical reaction at the atomic level without using chemicals at all, only quantum computers. This new branch of science, computational chemistry, determines chemical properties not by experiment, but by simulating them in a quantum computer, which may one day eliminate expensive and time-consuming testing. All of biology, medicine, and chemistry would be reduced to quantum mechanics. This means creating a "virtual laboratory" in which we can rapidly try out new drugs, therapies, and cures in the memory of a quantum computer, bypassing decades of trial and error and slow, tedious laboratory experiments. Instead of performing thousands of complex, expensive, and time-consuming chemical experiments, one might simply push the button on a quantum computer.

4. Merger of AI and Quantum Computers

Artificial intelligence (AI) excels at being able to learn from mistakes, so that it can perform increasingly difficult tasks. It has already proven its worth in industry and medicine. However, one limitation of AI is that the vast amount of data that it must process can easily overwhelm a conventional digital computer. But the ability to sift through mountains of data is one of the strong points of quantum computers. So the cross-fertilization of AI and quantum computers can significantly increase their power to solve problems of all kinds.

Further Applications of Quantum Computers

Quantum computers have the power to change entire industries. For example, quantum computers may finally usher in the long-awaited Solar Age. For decades, futurists and visionaries have predicted that renewable energy would phase out fossil fuels and solve the greenhouse effect that is warming our planet. Armies of these thinkers and dreamers have extolled the virtues of renewable energy.

But the Solar Age got sidetracked.

While costs have dropped for wind turbines and solar panels, they still only represent a small fraction of the world's energy production. The question is: What happened?

Every new technology has to confront the bottom line: costs. After decades of singing the praises of solar and wind power, boosters have to face the fact that it is still a bit more expensive than fossil fuels on average. The reason is clear. When the sun does not shine and the winds don't blow, renewable energy technology sits there unused, gathering dust.

The key bottleneck for the Solar Age is often overlooked; it is the battery. We have been spoiled by the fact that computer power grows exponentially fast, and we unconsciously assume that the same pace of improvement applies for all electronic technology.

Computer power has exploded in part because we can use shorter wavelengths of ultraviolet radiation to etch tiny transistors on a silicon chip. But batteries are different; they are messy, using a collection of exotic chemicals in complex interplay. Battery power grows slowly and tediously, by trial and error, not by systematically etching with shorter wavelengths of UV light. Furthermore, the energy stored in a battery is a tiny fraction of the energy stored in gasoline.

Quantum computers could change that. They may be able to model thousands of possible chemical reactions without having to perform them in the laboratory in order to find the most efficient process for a super battery, thereby ushering in the Solar Age.

Already, utilities and car companies are using first-generation quantum computers from IBM to attack the battery problem. They are trying to increase the capacity and recharging speed for the next generation of lithium-sulfur batteries. But this is just one way climate will be affected. In addition, ExxonMobil is using IBM's quantum computers to create new chemicals for low-energy processing and carbon capture. In particular, they want quantum computers to be able to simulate materials and determine their chemical nature, such as heat capacity.

Jeremy O'Brien, founder of PsiQuantum, emphasizes that this revolution is not about building faster computers. Instead, it's about tackling problems, like complex chemical and biological reactions, that no conventional computer could solve no matter how much time we gave it.

He says, "We're not talking about doing things faster or better . . . we're talking about being able to do these things at all. . . . These problems are forever beyond the reach of any conventional computer that we could ever build . . . even if we took every silicon atom on the planet and turned it into a supercomputer, we still could not solve these . . . hard problems."

Feeding the Planet

Another crucial application of quantum computers might be to feed the world's growing population. Certain bacteria can effortlessly take nitrogen from the air and convert it into ammonia, which is then turned into chemicals that become fertilizer. This nitrogen-fixing process is the reason why life flourishes on earth, allowing for the growth of lush vegetation that feeds humans and animals. The Green Revolution was unleashed when chemists duplicated this feat with the Haber-Bosch process. However, this process requires a vast amount of energy. In fact, an astounding 2 percent of the entire energy production of the world goes into this process.

So that is the irony. *Bacteria can do something for free that consumes a huge fraction of the world's energy.*

The question is: Can quantum computers solve this problem of efficient fertilizer production, creating a second Green Revolution? Without another revolution in food production, some futurists have predicted an ecological catastrophe as an ever-expanding world population becomes more and more difficult to feed, which could lead to mass starvation and food riots around the globe.

Already, scientists at Microsoft have made some of the first attempts to use quantum computers to increase the yields from fertilizers and unlock the secret of nitrogen fixing. In the end, quantum computers may help save human civilization from itself. Yet another miracle of nature is photosynthesis, in which sunlight and carbon dioxide are turned into oxygen and glucose, which then forms the foundation of nearly all animal life. Without photosynthesis, the food chain collapses and life on this planet would quickly wither away.

Scientists have spent decades trying to tease apart all the steps behind this process, molecule for molecule. But the problem of converting light into sugar is a quantum mechanical process. After years of effort, scientists have isolated where quantum effects dominate this process, and all are beyond the reach of digital computers. Therefore, to create a synthetic photosynthesis that could potentially be more efficient than the natural one still eludes our finest chemists.

Quantum computers may be able to help create a more efficient synthetic photosynthesis or perhaps entirely new ways of capturing the power of sunlight. The future of our food supply may depend on this.

Birth of Quantum Medicine

So quantum computers have the power to rejuvenate the environment and plant life. But they can also heal the sick and dying. Not only can quantum computers simultaneously analyze the efficacy of millions of potential drugs faster than any conventional computer, they can also unravel the nature of disease itself.

Quantum computers may answer questions like: What causes healthy cells to suddenly become cancerous, and how can they be stopped? What causes Alzheimer's disease? Why are Parkinson's and

ALS incurable? More recently, the coronavirus has been known to mutate, but how dangerous are each of these mutant viruses and how will they respond to treatment?

Two of the greatest discoveries in all of medicine are antibiotics and vaccines. But new antibiotics are found largely by trial and error, without understanding precisely how they work at the molecular level, and vaccines only stimulate the human body to produce chemicals to attack an invading virus. In both cases, the precise molecular mechanisms are still a mystery, and quantum computers may offer insight into how we might develop better vaccines and antibiotics.

When it comes to understanding the body, the first giant step was the Human Genome Project, which listed all of the 3 billion base pairs and 20,000 genes that form a blueprint for the human body. But this is just the beginning. The problem is that digital computers are used mainly to search through vast databases of known genetic codes, but they are helpless when it comes to explaining precisely how DNA and proteins perform their miracles inside the body. Proteins are complex objects, often consisting of thousands of atoms, which fold up into a small ball in specific and unexplainable ways when they do their molecular magic. At its most fundamental level, all life is quantum mechanical, and so beyond the reach of digital computers.

But quantum computers will lead the way into the next stage, when we decipher the mechanisms at the molecular level that tell us how they work, allowing scientists to create new genetic pathways, new therapies, new cures to conquer previously incurable diseases.

For example, pharmaceutical corporations, including firms like ProteinQure, Digital Health 150, Merck, and Biogen, are already setting up research centers to analyze how quantum computers will affect drug analysis.

Scientists are amazed that Mother Nature has been able to create a vast arsenal of molecular mechanisms that make the miracle of life possible. But these mechanisms are a by-product of chance and random natural selection operating over billions of years. That is why we still suffer from certain incurable diseases and the aging process. Once we understand how these molecular mechanisms work, then we will be able to use quantum computers to improve on them or create new versions of them.

For example, with DNA genomics, we can use computers to identify genes like BRCA1 and BRCA2 that can likely result in breast cancer. But digital computers are useless to determine precisely how these defective genes cause cancer. And they are also powerless to stop cancer once it spreads throughout the body. But quantum computers, by deciphering the molecular intricacies of our immune system, may be able to create new drugs and therapies to combat these diseases.

Another example is Alzheimer's disease, which some believe will be the "disease of the century" as the world population ages. With digital computers, one can show that mutations in certain genes, like the ApoE4 gene, are associated with Alzheimer's. But digital computers are useless in explaining why this is so.

One leading theory is that Alzheimer's is caused by prions, a certain amyloid protein that is incorrectly folded in the brain. When the renegade molecule bumps into another protein molecule, it causes that molecule to fold up the wrong way as well. Thus, the disease can spread by contact, even though bacteria and viruses are not involved. It is suspected that renegade prions might be the culprit behind Alzheimer's, Parkinson's, ALS, and a host of other incurable diseases that target the elderly.

So the protein folding problem is one of the greatest, uncharted areas in biology. In fact it may hold the secret of life itself. But precisely how a protein molecule folds up is beyond the capability of any conventional computer. Quantum computers, however, can provide new pathways by which to neutralize renegade proteins and provide new therapies.

In addition, the aforementioned merger of AI and quantum computers may turn out to be the future of medicine. Already, AI programs like AlphaFold have been able to map the detailed atomic structure of an astounding 350,000 different types of proteins, including the complete set of proteins that make up the human body. The next step is to use the unique methods of quantum computers to find out how these proteins do their magic, and to use them to create the next generation of drugs and therapies.

Quantum computers are already being connected to neural networks, to create the next generation of learning machines that can literally reinvent themselves. The laptop sitting on your desk, by contrast, never learns. It is no more powerful today than it was last year. Only recently, with new advances in deep learning, are computers taking the first steps to recognizing mistakes and learning. Quantum computers could exponentially accelerate this process and have singular impacts on medicine and biology.

Google CEO Sundar Pichai compares the arrival of quantum computers to the Wright brothers' historic 1903 flight. The original test was not so amazing by itself, because the flight lasted only a modest twelve seconds. But this short flight was the trigger that initiated modern aviation, which in turn has changed the course of human civilization.

What is at stake is nothing less than our future. It's up for grabs for whoever is able to build and use a quantum computer. But to truly understand the impact this revolution might have on our daily lives, it is useful to retrace some of the valiant attempts made in the past to fulfill our dream of using computers to simulate and understand the world around us.

And it all began with a mysterious, 2,000-year-old relic found at the bottom of the Mediterranean.

CHAPTER 2

END OF THE DIGITAL AGE

f rom the bottom of the Aegean Sea came one of the most intriguing, captivating puzzles of the ancient world. In 1901, divers were able to salvage a strange curiosity near the island of Antikythera. Among the scattered pieces of broken pottery, coins, jewelry, and statues in a shipwreck, divers found one object that was oddly different. At first, it looked like a worthless piece of coral-encrusted rock.

But when layers of debris were cleaned off, archaeologists began to realize that they were staring at an exceedingly rare, one-of-a-kind treasure. It was full of gears, wheels, and strange inscriptions, a machine of intricate and exquisite design.

Dating the artifacts found within the shipwreck, it was estimated that it was crafted sometime between 150 and 100 BCE. Some historians believe it was being taken from Rhodes to Rome, to be given as a gift to Julius Caesar for a triumphal parade.

In 2008, scientists using X-ray tomography and high-resolution surface scanning were able to penetrate the interior of this intriguing object. They were shocked when they realized that they were staring at an ancient mechanical device that was unbelievably advanced.

Nowhere in the ancient record was there any mention of a mechanism this sophisticated. It dawned on them that this magnificent machine must have been the pinnacle of scientific knowledge of the ancient world. It was a supernova of brilliance staring at them from millennia past. This was the *world's oldest computer,* a device that would not be duplicated for another two thousand years.

Figure 1: The Antikythera mechanism
Two thousand years ago, the Greeks created the Antikythera, the very first in the long evolutionary line of computers, depicted here as a model based on the original device. While the Antikythera represents the beginning of computer technology, the quantum computer may represent the highest stage in its evolution.

Scientists began to construct mechanical reproductions of this remarkable device. By turning a crank, a series of complex wheels and cogs were set into motion for the first time in thousands of years. It had at least thirty-seven bronze gears. In one set of gears, the motion of the moon and sun were calculated. Another set of gears could predict the coming of the next eclipse of the sun. It was so sensitive it could even calculate small irregularities in the orbit of the moon. Translations of the inscriptions on the device chronicle the motion of Mercury, Venus, Mars, Saturn, and Jupiter, the planets known to the ancients, but it is believed that yet another portion of the device, which is missing, could actually plot out the planets as they move in the heavens.

Since then, scientists have created elaborate models of the interior of the device, which have given historians unprecedented insight into the knowledge and mind of the ancients. The device heralded the birth of an entirely new branch of science, which uses mechanical tools to simulate the universe. This was the world's oldest analog computer—a device that could calculate using continuous mechanical motions.

So the purpose of the world's first computer was to simulate heavenly bodies, to reproduce the mysteries of the cosmos in a device you could hold in your hands. Instead of just staring in awe at the night sky, these ancient scientists wanted to understand its detailed workings, allowing them unprecedented insight into the motion of celestial bodies in the heavens.

Quantum Computers: The Ultimate Simulation

Archaeologists found that the Antikythera represented the pinnacle of our ancient attempts to simulate the cosmos. In fact, this same age-old urge to simulate the world around us is one of the driving forces behind the quantum computer, which represents the ultimate effort in the 2,000-year journey to simulate everything from the cosmos down to the atom itself.

Simulation is one of our deepest human desires. Children use simulation with toy figures to understand human behavior. When children play cops and robbers, teacher and student, or doctor and patient, they are simulating a piece of adult society in order to understand complex human relations.

Sadly, it would take many centuries before scientists could build machines of sufficient complexity to simulate our world as well as the Antikythera could.

Babbage and the Difference Machine

With the fall of the Roman Empire, scientific progress in many areas, including simulating the universe, came to a standstill.

It wasn't until the 1800s that interest was gradually revived. By then, there were urgent, practical questions that could only be answered by mechanical analog computers.

For example, navigators depended on detailed maps and charts to plot the courses of their ships. They needed devices to help make these maps as accurate as possible.

Machines of increasing complexity were also needed to keep track of trade and commerce as people began to accumulate wealth in greater and greater quantities. Accountants were required to compile large mathematical tables of interest and mortgage rates by hand.

But humans, however, would often make costly and crucial errors. So there was acute interest in devising mechanical adding machines that wouldn't make these mistakes. As adding machines became more complex, an informal competition arose among enterprising inventors to see who could build the most advanced one.

Perhaps the most ambitious of these projects was led by the eccentric English inventor and visionary Charles Babbage, who is often called the Father of the Computer. He dabbled in a number of disparate fields, including art and even politics, but was always fascinated by numbers. Fortunately, he was born into a wealthy family, so his banker father could help him pursue many of his diverse interests.

His dream was to create the most advanced computing machine of his time, which could be used by bankers, engineers, sailors, and the military to unerringly perform tedious but essential calculations. He had two goals. As a founding member of the Royal Astronomical Society, he had an interest in creating a machine that could track the motion of the planets and astronomical bodies (essentially following in the same pioneering path taken by those who built the

Antikythera). He was also concerned with producing accurate navigational charts for the maritime industry. England was a major seafaring power, and errors in navigational charts could cause expensive disasters. His idea was to create the most powerful mechanical computer of its kind to plot the motion of everything from the planets to ships at sea to interest rates.

He was quite persuasive in recruiting eager followers to help him advance his ambitious project. One of them was Lady Ada Lovelace, a member of the aristocracy and daughter of Lord Byron. She was also a serious student of mathematics, which was rare for women of that time. When she saw a small working model of Babbage's project, she became intrigued by this exciting program.

Lovelace is known for helping Babbage introduce several new concepts in computing. Usually, a mechanical computer required a set of gears and cogs to slowly and painstakingly calculate numbers, one by one. But to generate entire tables full of thousands of mathematical numbers at one time (such as logarithms, interest rates, and navigation charts), one needed a set of instructions to guide the machine across many iterations. In other words, one needed software to guide the sequence of computations in the hardware. So she wrote a series of detailed instructions by which the machine could systematically generate what are called Bernoulli numbers, essential for the calculations it performed.

Lovelace was, in a sense, the world's first programmer. Historians agree that Babbage was probably aware of the importance of software and programming, but her detailed notes written up in 1843 represented the first published account of a computer program.

She also recognized that the computer was not just capable of manipulating numbers, as Babbage thought, but could be generalized to describe symbolic concepts over a wide range of areas as well. Author Doron Swade writes, "Ada saw something that Babbage in some sense failed to see. In Babbage's world his engines were bound by number. What Lovelace saw . . . was that number could represent entities other than quantity. So once you had a machine for manipulating numbers, if those numbers represented other things, letters, musical notes, then the machine could manipulate symbols of which number was one instance, according to rules."

As an example, Lovelace wrote that the computer could be programmed to create musical pieces. She wrote, "the engine might compose elaborate and scientific pieces of music of any degree of complexity or extent." So the computer was not just a number cruncher or glorified adding machine. It could be used to explore science, art, music, and culture as well. But unfortunately, before she could elaborate on these world-changing concepts, she died of cancer at the age of thirty-six.

Meanwhile, because Babbage was chronically out of funds and continually engaged in disputes with others, his dream to create the most advanced mechanical computer of its time was never finished. When he died, many of his blueprints and ideas died with him.

But since then, scientists have tried to study precisely how advanced his machines were. The blueprint of one of his unfinished models contained 25,000 parts. When built, it would have weighed four tons and would have been eight feet tall. He was so far ahead of his time that his machine could have manipulated one thousand 50-digit numbers. That huge amount of memory would not be duplicated by another machine until 1960.

But about a century after his death, engineers at the London Science Museum, by following his designs on paper, were able to finish one of his models and put it on display. And it worked, just as Babbage had predicted in the previous century.

Is Mathematics Complete?

While engineers were building ever more complex mechanical computers to meet the demands of an increasingly industrial world, pure mathematicians were asking yet another question. It had always been one of the dreams of the Greek geometers to show that all true statements in mathematics could be rigorously proven.

But remarkably, this simple idea frustrated mathematicians for 2,000 years. For centuries, students of Euclid's *Elements* would struggle to prove theorem after theorem about geometric objects. Over time, brilliant thinkers were able to prove an increasingly elaborate set of true statements. Even today, mathematicians spend their entire lives compiling scores of true statements that can be proven mathematically. But in Babbage's day they began to ask an even more fundamental question: Is mathematics complete? Do the rules of mathematics ensure that every true statement can be proven, or are there true statements that can elude the most exceptional minds of the human race because they are, in fact, not provable?

In 1900, the great German mathematician David Hilbert listed the most important unproven mathematical questions of the time, challenging the world's greatest mathematicians. This remarkable set of unsolved questions would then guide the agenda of mathematics for the next century as, one by one, each unproven theorem would be proven. Over the decades, young mathematicians would find fame and glory as they conquered one of Hilbert's unfinished theorems.

But there was some irony here. One of the unsolved problems listed by Hilbert was the ancient problem of proving all true statements in mathematics when given a set of axioms. In 1931, at a conference where Hilbert was discussing his program, a young Austrian mathematician, Kurt Gödel, proved it was impossible.

Shock waves rippled through the mathematics community. Two thousand years of Greek thinking was completely and irrevocably shattered. Mathematicians around the world were left shaking their heads in utter disbelief. They had to come to grips with the fact that mathematics was not the neat, tidy, complete, and provable set of theorems once postulated by the Greeks. Even mathematics, which formed the foundation for understanding the physical world around us, was messy and incomplete.

Alan Turing: Computer Science Pioneer

A few years later, one young English mathematician, who was intrigued by Gödel's famous incompleteness theorem, found an ingenious way to reframe the entire question. It would forever change the direction of computer science.

Alan Turing's exceptional ability was recognized early in his life. The headmistress of his elementary school would write that, among her students, she "has clever boys and hardworking boys, but Alan is a genius." He would later be known as the father of computer science and artificial intelligence.

Turing had a fierce determination to master mathematics in spite of harsh opposition and hardship. His headmaster, in fact, would actively try to discourage his interest in science, stating that "he is wasting his time at a public school." But this opposition only further inflamed his determination. When he was fourteen years old, there was a general strike that shut down much of the country, but he was so eager to be at school that he rode sixty miles on a bike by himself to be in class when it opened again.

Instead of building increasingly complex adding machines like Babbage's difference engine, Alan Turing eventually asked himself a different question: Is there a mathematical limit to what a mechanical computer can perform?

In other words, can a computer prove everything?

To do this, he had to make the field of computer science rigorous, since it was previously a loose collection of disjointed ideas and inventions by eccentric engineers. There was no systematic way in which to discuss questions like the limit of what is computable. So in 1936 he introduced the concept of what is now called a universal Turing machine, a deceptively simple device that captured the essence of computation, allowing the entire field to be put on a firm mathematical basis. Today, Turing machines are the foundation for all modern computers. Everything, from the giant supercomputers of the Pentagon to the cell phone in your pocket, are all examples of Turing machines. It is no exaggeration to say that almost all of modern society is built on Turing machines.

Turing imagined an infinitely long tape, which contained a series of squares or cells. Inside each square, you could put a 0 or a 1, or you could leave it blank.

Then a processor read the tape, and was allowed to make just six simple operations on it. Basically, you could replace a 0 with a 1, or vice versa, and move the processor one square to the left or right:

1. You can read the number in the square
2. You can write a number in the square
3. You can move one square to the left
4. You can move one square to the right
5. You can change the number in the square
6. You can stop

(The Turing machine is written in binary language, rather than base 10. In binary language, the number one is represented by 1, the number two is represented by 10, the number three is represented by 11, the number four by 100, and so on. There is also a memory where numbers can be stored.) Then the final numerical result emerges from the processor as output.

In other words, the Turing machine can take one number and turn it into another according to precise commands in the software. So Turing reduced mathematics to a game: by systematically replacing 0 with 1 and vice versa, one could encode all of mathematics.

In the paper that laid out these ideas, Turing showed with a concise set of instructions that one could use his machine to perform all the manipulations of arithmetic, i.e., it could add, subtract, multiply, and divide. He then used this result to prove some of the most diffi-

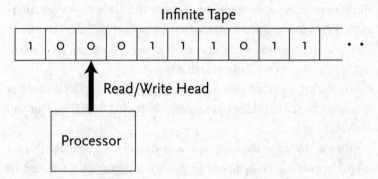

Figure 2: Turing machine
A Turing machine consists of (a) an infinitely long input digital tape, (b) an output digital tape, and (c) a processor that converts the input information into the output according to a fixed set of rules. It is the basis of all modern digital computers.

cult problems in mathematics, rephrasing everything from the point of view of computability. The sum total of all of mathematics was being rewritten from the point of view of computation.

For example, let us show how 2 + 2 = 4 is done on a Turing machine, which demonstrates how all of arithmetic may be encoded. Start the tape with the input given by the number two, or 010. Then move to the middle cell, where there is a 1, and replace it with 0. Then move one step the left, where there is a 0 and replace that with 1. The tape now reads 100, which is equal to four. By generalizing these commands, one can perform any operation involving addition, subtraction, and multiplication. With a little more work one can also divide numbers as well.

Turing then asked himself a simple but important question: Can Gödel's infamous incompleteness theorem, which involved higher mathematics, be proven using his Turing machine, which was much simpler but still captured the essence of mathematics?

Turing began by defining what is computable. He said, in essence, that a theorem is computable if it can be proven in a finite amount of time by a Turing machine. If a theorem requires an infinite amount of time on a Turing machine, then, for all intents and purposes, the theorem is not computable, and we don't know if the theorem is correct or not. Therefore, it would not be provable.

Simply put, Turing then expressed the question raised by Gödel in a concise form: Are there true statements that cannot be computed in a finite amount of time by a Turing machine, given a set of axioms?

Like the work of Gödel, Turing showed that the answer is yes.

Once again, this shattered the ancient dream of proving the completeness of mathematics, but in a way that was intuitive and simple. It meant that, even with the most powerful computer in the world, one can never prove all true statements in mathematics in a finite amount of time with a given set of axioms.

Computers in Warfare

Clearly, Turing had proved himself to be a mathematical genius of the highest caliber. But his research was interrupted by World War II. To aid in the war effort, Turing was recruited to perform top secret work at the British military installation at Bletchley Park outside London. There they were tasked with decoding secret Nazi codes. Nazi scientists had created a machine, called Enigma, which could take a message, rewrite it with an unbreakable code, and then send the scrambled message to the globe-spanning Nazi war machine. Within the code were the most sensitive set of instructions in the world: the war plans of the Nazi military, particularly the navy. The ultimate fate of civilization might depend on cracking the Enigma code.

Turing and his colleagues set upon this crucial problem by designing calculating machines that might systematically crack these impenetrable codes. Their first breakthrough, called the bombe, resembled Babbage's difference engine in some ways. Instead of steam-driven mechanisms like previous machines, whose gears and cogs were slow, difficult to make, and often jammed, the bombe relied on rotors, drums, and relays, all powered by electricity.

But Turing was also involved with another project, Colossus, with an even more ingenious design. Historians believe it was the world's first *programmable digital electronic computer*. Instead of mechanical parts like the difference engine or the bombe, they used vacuum tubes, which can send electrical signals near the speed of light. Vacuum tubes can be compared to valves controlling the flow of water. By turning a small valve, one can shut off the water flowing in a much larger pipe, or let it flow unimpeded. This, in turn, can represent the number 0 or 1. So a system of water pipes and valves can represent a digital computer, where the water is like the flow of electricity. In the machines at Bletchley Park, a large array of vacuum tubes could perform digital calculations at enormous speeds by turning the flow of electricity on or off in the vacuum tubes. Thus, the work of Turing and others replaced the analog computer with a digital computer. One version of Colossus contained 2,400 vacuum tubes and filled up an entire room.

Besides being faster, digital computers have another great advantage over analog systems. Think of using an office copy machine to repeatedly duplicate a picture. Each time you recycle the picture by copying it, you lose some information. If you recycle the same image over and over again, eventually the image gets fainter and fainter and finally disappears completely. So analog signals are prone to introduce errors each time the image is copied.

(Now instead, digitize the picture so it becomes a series of 0s and 1s. When you first digitize the picture, you will lose some information. However, a digital message can be copied over and over again, and lose hardly any information with each cycle. So digital computers can be vastly more accurate than analog computers.

(Also, it's easy to edit digital signals. Analog signals, like a picture, are exceedingly difficult to alter. But digital signals can be altered with the push of a button by using simple mathematical algorithms.)

Under the enormous pressure of wartime, Turing and his team were able to finally break the Nazi code around 1942, which helped defeat the Nazi naval fleet in the Atlantic. Soon, the Allies were able to penetrate the deepest secret plans of the Nazi military. The Allies were able to eavesdrop on Nazi instructions to their troops and anticipate their war plans. Colossus was finished in 1944, in time for the final invasion of Normandy, which the Nazis did not adequately prepare for. This sealed the fate of the Nazi empire.

These were breakthroughs of monumental proportions, some of which were immortalized in the 2014 movie *The Imitation Game*. Without their key achievements the war might have dragged on for years, creating untold misery and suffering. Historians such as Harry Hinsley have estimated that the work of Turing and others at Bletchley Park shortened the length of war by about two years and saved over 14 million lives. The map of the world, and the lives of untold numbers of innocents, were irrevocably altered by his pioneering work.

In the U.S., the workers who built the atomic bomb were heralded as war heroes and as miracle workers, but a different fate awaited Turing in the U.K. Because of national secrecy laws, his accomplishments were kept classified for decades, so no one knew of his enormous contribution to the war effort.

Turing and the Creation of AI

After the war, Turing returned to an age-old problem that had intrigued him as a youth: artificial intelligence. In 1950, he opened his landmark paper on the subject by stating, "I propose to consider the question: Can machines think?"

Or to put it another way, is the brain a Turing machine of some sort?

He was tired of all the philosophical discussions that stretched back centuries about the meaning of consciousness, the soul, and what makes us human. Ultimately, all this discussion was pointless, he thought, because there was no definitive test or benchmark for consciousness.

So Turing came up with the celebrated Turing test. Put a human in a sealed room and a robot in another room. You are allowed to ask each one any written question and read their responses. The challenge is: Can you determine which room held the human? He called this test the imitation game.

In his paper, he wrote, "I believe that in about fifty years' time, it will be possible to programme computers, with a storage capacity of about 10^9, to make them play the imitation game so well that an average interrogator will not have more than 70 percent chance of making the right identification after five minutes of questioning."

The Turing test replaces endless philosophical debate with a simple reproducible test, to which there is a simple yes-or-no answer. Unlike a philosophical question, for which there is often no answer, this test is decidable.

Furthermore, it sidesteps the slippery question of "thinking" by simply comparing it to whatever humans can do. There is no need to define what we mean by "consciousness," "thinking," or "intelligence." In other words, if something looks and acts like a duck, then maybe it *is* a duck, regardless of how you define it. He gave an operational definition of intelligence.

So far, no machine has been able to consistently pass the Turing test. Every few years, headlines are made when the Turing test is conducted, but each time the judges are able to tell the difference between a human and a machine, even if it is allowed to tell lies and make up facts.

But an unfortunate incident would put an abrupt end to all of Turing's pathbreaking work.

In 1952, someone burglarized Turing's home. When the police came to investigate, they found evidence that Turing was gay. For this, he was arrested and sentenced under the Criminal Law Amendment Act of 1885. The punishment was quite harsh. He was given a choice of going to prison or undergoing a hormonal procedure. When he chose the latter, he was given stilboestrol, a synthetic form of the female sex hormone estrogen, which caused him to grow breasts and become impotent. The controversial treatments lasted for one year. Then one day, he was found dead in his home. He died from a fatal dose of cyanide poison. It was reported that next to him there was a half-eaten poison apple, which some speculated was how he committed suicide.

It is a tragedy that one of the creators of the computer revolution, who helped save the lives of millions and defeat fascism, was in some sense destroyed by his own country.

But his legacy lives on in every digital computer on the planet. Today, every computer on earth owes its architecture to the Turing machine. The world economy depends on the pioneering work of this man.

But this is only the beginning of our story. Turing's work is based on something called determinism, i.e., the idea the future is determined ahead of time. This means that if you fed a problem into a Turing machine, you would get the same answer every time. Everything in this sense is predictable.

Therefore if the universe were a Turing machine, all future events would have been determined at the instant the universe was born.

But another revolution in our understanding of the world would overturn this idea. Determinism would be overthrown. In the same way that Gödel and Turing helped show that mathematics is incomplete, perhaps computers of the future would have to deal with the fundamental uncertainty introduced by physics.

So mathematicians would focus on a different question: Is it possible to build a quantum Turing machine?

CHAPTER 3

RISE OF THE QUANTUM

ax Planck, the creator of the quantum theory, was a man of many contradictions. On one hand, he was the ultimate conservative. It might be because his father was a professor of law at the University of Kiel, and his family had a long, distinguished tradition of integrity and public service. Both his grandfather and great-grandfather were theology professors, and one of his uncles was a judge.

He was cautious in his work, ever precise in his manners, and a pillar of the establishment. Judging by appearances, this mild-mannered man would be the last person you would think would become one of the greatest revolutionaries of all time, shattering all the cherished notions of previous centuries by opening up the quantum floodgates. But that is exactly what he did.

In the year 1900, leading physicists were firmly convinced that the world around us could be fully explained by the work of Isaac Newton, whose laws described the motions of the universe, and James Clerk Maxwell, who discovered the laws of light and electromagnetism. Everything, from the motion of giant planets in space to cannon balls to lightning bolts, could be explained by Newton and Maxwell. It was said that the U.S. Patent Office even contemplated shutting down because everything that could be invented already had been.

According to Newton, the universe was a clock. It was ticking away following his three laws of motion in a precise and predetermined way. This was called Newtonian determinism, which held sway for several centuries. (It is sometimes called classical physics, to distinguish it from quantum physics.)

But there was a nagging problem. There were a few loose strings, and by pulling on them, this elaborate Newtonian architecture would eventually unravel.

Ancient artisans knew that if clay were heated up to high enough temperature in a furnace, it would eventually glow brightly. It would start becoming red hot, then yellow hot, and finally bluish-white hot. We see this every time we light up a match. At the top of the flame, where it is coolest, the flame is red. In the center, the flame is yellow. And, if the conditions are right, the bottom of the flame is blue-white hot.

Physicists tried to derive this well-known property of hot objects and failed miserably. They knew that heat is nothing but atoms in motion. The greater the temperature of an object, the faster its atoms move. They also knew that atoms have electrical charges as well. If you move a charged atom fast enough, it radiates electromagnetic radiation (like radio or light), according to the laws of James Clerk Maxwell. The color of a hot object indicates the frequency of the radiation.

So, using Newton's theory applied to the atom, and using Maxwell's theory of light, one can calculate the light emitted from a hot object. So far, so good.

But when the calculation is actually performed, disaster strikes. One finds that the energy emitted can become infinite at high frequencies, which is impossible. This was called the Rayleigh-Jeans catastrophe. It showed physicists that there was a gaping hole in Newtonian mechanics.

One day, Planck tried to derive the Rayleigh-Jeans catastrophe for his physics class, but with a strange, novel method. He was tired of doing it in the same old-fashioned way, so, for purely pedagogical reasons, he made a bizarre assumption. He supposed that the energy emitted from an atom could only be found in tiny discrete packs of energy, which are called quanta. This was heresy, because Newton's equations stated that energy should be continuous, not in packets. But when Planck postulated that the energy occurred in packets of a certain size, he found precisely the correct curve linking frequency and energy for light.

This was a discovery for the ages.

Birth of the Quantum Theory

It was the first step in a long process that would eventually create the quantum computer.

Planck's revolutionary insight meant that Newtonian mechanics was incomplete, and a new physics must emerge. Everything we thought we knew about the universe would have to be completely rewritten.

But being a proper conservative, he proposed his idea cautiously, diplomatically claiming that if you introduce this trick of packets of energy as an exercise, then you can precisely reproduce the actual energy curve found in nature.

To do the calculation, he had to introduce a number representing the size of the quantum of energy. He called it h (otherwise known as Planck's constant, $6.62 \ldots \times 10^{-34}$ joule-seconds), which is an incredibly small number. In our world, we never see quantum effects because h is so small. But if you could somehow vary h, one could continuously move from the quantum world to our everyday world. Almost like tuning a radio dial, one could turn it all the way down, so $h = 0$, and we have the commonsense world of Newton, where there are no quantum effects. But turn it the other way, and we have the bizarre subatomic world of the quantum, a world that, as physicists would shortly find out, resembled the Twilight Zone.

We can also apply this to a computer. If we let h go to zero, we arrive at the classical Turing machine. But if we let h get larger, then quantum effects begin to emerge, and we slowly turn the classical Turing machine into a quantum computer.

Although his theory indisputably fit the experimental data and opened up an entirely new branch of physics, he was hounded for years by stubborn, die-hard believers in the classical, Newtonian idea. Describing this blizzard of opposition, Planck wrote: "A new scientific truth does not triumph by convincing its opponents and making them see the light, but rather because its opponents eventually die, and a new generation grows up that is familiar with it."

But no matter how fierce the opposition was, more and more evidence began to pile up confirming the quantum theory. It was indisputably correct.

For example, light, when it hits a metal, can knock out an electron, which creates a small electrical current, known as the photoelectric effect. This is what allows a solar panel to absorb light and convert it into electricity. (It is also commonly used in many appliances, such as solar-powered calculators, which replace batteries with solar cells, and modern digital cameras, converting light from the subject into electrical signals.)

The man who finally explained this effect was a penniless, unknown physicist, toiling away in an obscure patent office in Berne, Switzerland. As a student, he cut so many classes that his professors wrote him unflattering letters of recommendation, resulting in his being rejected for every teaching position he applied for after graduation. He was often unemployed and bounced around in a string of odd jobs, such as tutor and salesman. He even wrote a letter to his parents that perhaps it would have been better if he had never been born. Finally, he wound up as a low-level clerk at the patent office. Most people would call him a failure.

The man who explained the photoelectric effect was Albert Einstein, and he did it using Planck's theory. Following Planck, Einstein claimed that light energy could occur in discrete packets or quanta of energy (later called photons) that could knock electrons out of a metal.

Thus, a new physical principle began to emerge. Einstein introduced the concept of "duality," i.e., that light energy has a dual nature. Light could act like a particle, the photon, or a wave, as in optics. Somehow, light had two possible forms.

In 1924, a young grad student, Louis de Broglie, using the ideas of Planck and Einstein, made the next big leap. If light can occur both as a particle and a wave, then why not matter? Perhaps electrons also possessed duality.

This was heretical, since matter was believed to be made of particles, called atoms, an idea introduced by Democritus 2,000 years earlier. But eventually a clever experiment overturned this belief.

When you throw rocks into a pond, ripples are formed that expand and then bump into each other, thereby creating a weblike interference pattern on the surface of the pond. This explains the properties of waves, but matter was thought to be based on pointlike particles, which have no wavelike interference pattern.

But now start with two parallel sheets of paper. In the first sheet, cut two small slits, and shine a light beam through the slits. Because light has wavelike properties, a distinct pattern of light and dark bands emerges on the second sheet. As the waves pass through both slits they interfere with each other on the second sheet, amplifying and canceling each other out, creating bands called interference patterns. This was well known.

But now, modify this experiment by replacing the light beam with an electron beam and the second sheet of paper with photo paper. If a beam of electrons was shot through two slits in the first sheet of paper, then one would expect there to be two distinct bright slits on the other sheet. This is because the electron was thought to be a point particle, and would pass either through the first slit or the second, but not both.

When this experiment was replicated with electrons, researchers found a wavelike pattern, similar to the effect of the beam of light. Electrons were acting as if they were waves, not just point particles. Atoms were long thought to be the ultimate unit of matter. Now, they were dissolving into waves like light. These experiments demonstrated that atoms could behave like either a wave or a particle.

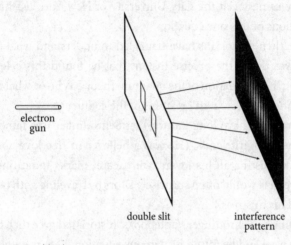

Figure 3: Double slit experiment
If a beam of electrons hits a barrier with two slits, instead of forming an image with two distinct slits, it forms a complex wavelike interference pattern. This is also true if only one electron is sent through. In some sense, a single electron has traveled through both holes. Even today, physicists debate how an electron can be in two places at the same time.

One day, Austrian physicist Erwin Schrödinger was discussing the idea of matter as a wave with a colleague. But if matter can act like a wave, his friend asked, then what is the equation that it must obey?

Schrödinger was intrigued by that question. Physicists were familiar with waves, since they were useful in studying the optical properties of light, and were often analyzed in the form of ocean waves or sound waves in music. So Schrödinger set out to find the wave equation for electrons. It was an equation that would completely overturn our understanding of the universe. In some sense, the entire universe, with all its chemical elements, including you and me, are solutions of Schrödinger's wave equation.

Birth of the Wave Equation

Today, Schrödinger's wave equation is the bedrock of the quantum theory, taught in any graduate course in advanced physics. It forms the heart and soul of the quantum theory. I sometimes spend an entire semester at the City University of New York teaching the implications of this one equation.

Since then, historians have struggled to understand what Schrödinger was doing the precise instant that he found this celebrated equation, the foundation of the quantum theory. Who or what helped inspire one of the greatest creations of the century?

Biographers have long known that Schrödinger was famous for his numerous girlfriends. (He was a believer in free love and kept a notebook listing all his lovers, with secret marks indicating each encounter. He would often surprise visitors by traveling with both his wife and his mistress.)

Examining Schrödinger's notebooks, historians agree that, during the very weekend he found his famous equation, he was with one of his girlfriends in the Villa Herwig in the Alps. Some historians have called her the muse who inspired the quantum revolution.

Schrödinger's equation was a bombshell. It was an immediate, overwhelming success. Previously, physicists like Ernest Rutherford thought that the atom was like a solar system, with tiny pointlike electrons circling around a nucleus. This picture, however, was much too simplistic because it said nothing about the structure of the atom and why there were so many elements.

But if the electron was a wave, then the wave should form discrete resonances of definite frequencies as it circled around the nucleus. When one catalogued the resonances that an electron could make, one found a wave pattern that fit the description of the hydrogen atom perfectly.

How does this work? When we sing in the shower, only some of the waves from our voice can resonate between the walls, making a pleasing sound. We suddenly become great opera singers while in the shower. Other frequencies that don't fit correctly inside the shower eventually die out and fade away. Similarly, if we beat on a drum, or blow on a trumpet, only certain frequencies are allowed to vibrate on its surface or in its pipes. This is the basis of music.

By comparing the resonances predicted by Schrödinger's waves with actual elements, one found a remarkable one-to-one correspondence. Physicists, who for decades were stumped trying to understand the atom, were now able to peek inside the atom itself. When one compared these wave patterns to the hundred or so chemical elements found in nature by Dmitri Mendeleev and others, one could explain the chemical properties of the elements using pure mathematics.

This was a staggering achievement. Physicist Paul Dirac would write prophetically: "The fundamental laws necessary for the mathematical treatment of a large part of physics and the whole of chemistry are thus completely known, and the difficulty lies only in the fact that application of these laws leads to equations that are too complex to be solved."

The Quantum Atom

The periodic table of elements, which was painstakingly assembled by chemists over centuries, could now be explained using a simple equation, by solving for the resonances of electron waves as they whirl around the nucleus of the atom.

To see how the periodic table emerges from the Schrödinger equation, think of the atom as a hotel. Each floor has a different number of rooms, and each room can accommodate up to two electrons. Further, each room has to be filled up in a certain order, i.e., the first-floor room has to be occupied before the second floor can be booked. In the first floor, we have a room or "orbital" called 1S, which can accommodate one or two electrons. The 1S room corresponds to hydrogen in the case of one electron and helium in the case of two.

On the second floor, we have two types of rooms, called the 2S and 2P orbitals. In the 2S room, we can accommodate two electrons, but we also have three P rooms as well, which are labeled Px, Py, and Pz, and each has two electrons apiece. This means that we can have up to eight electrons on the second floor. These rooms, when filled, in turn correspond to lithium, beryllium, boron, carbon, nitrogen, oxygen, fluorine, and neon.

When an electron is not paired up in its room, it can be shared between different hotels that have available rooms. So, when two atoms come close to each other, the wave of one unpaired electron can be shared between atoms, so that the electron wave goes back and forth between the two. This creates a bond, forming a molecule.

The laws of chemistry can be explained as we fill up the hotel rooms. At the lowest level, if we have two electrons in the S orbital, then the 1S orbital is full. This means that helium, which has only two electrons, cannot form any chemical bonds, so it is chemically inert and does not form any molecules. Similarly, if we have eight electrons in the second level, then we fill up all the orbitals, so neon cannot form any molecules either. In this way, we can explain why there are inert gases like helium, neon, krypton.

This also helps explain the chemistry of life. The most important organic element is carbon, which has four bonds and hence can create hydrocarbons, which are the building blocks of life. Looking at the table, we see that carbon has four empty orbitals at the second level, which allows it to bond with four other atoms of oxygen, hydrogen, etc., to form proteins and even DNA. The molecules of our body are by-products of this simple fact.

The point is that, by determining how many electrons are in each level, one can simply and elegantly predict many of the chemical properties of the periodic table using pure mathematics. In this way, the entire periodic table can largely be predicted from first principles.

All the 100-plus elements of the table can roughly be described by the electrons in various resonances going around the nucleus, like filling up hotel rooms, floor by floor.

It was breathtaking to realize that a single equation could explain the elements that make up the entire universe, including life itself. Suddenly, the universe was simpler than anyone thought.

Chemistry has been reduced to physics.

Waves of Probability

As spectacular and powerful as the Schrödinger equation was, there was still one important, but embarrassing, question. If the electron was a wave, then what is waving?

The solution would divide the physics community right down the middle, pitting physicists against one another for decades to come. It would spark one of the most controversial debates in the entire history of science, challenging our very notion of existence. Even today, there are conferences debating all the mathematical nuances and philosophical implications of this split. And one by-product of this debate, as it would turn out, is the quantum computer.

Physicist Max Born lit the fuse of this explosion by postulating that matter consists of particles, *but the probability of finding that particle is given by a wave.*

This immediately cleaved the physics community in two, with the founders of the "old" guard on one side (including Planck, Einstein, de Broglie, and Schrödinger, all denouncing this new interpretation), and Werner Heisenberg and Niels Bohr on the other, creating the Copenhagen school of quantum mechanics.

This new interpretation was too much, even for Einstein. It meant that you can only calculate probabilities, never certainties. You never knew precisely where a particle was; you could only calculate the likelihood that it was there. In some sense, electrons can be two places at the same time. Werner Heisenberg, who came up with an alternative but equivalent formulation of quantum mechanics, would call this the uncertainty principle.

All of science was being turned upside down before their eyes. Previously, mathematicians were forced to confront the incompleteness theorem, and now physicists had to confront the uncertainty principle. Physics, like mathematics, was somehow incomplete.

So with this new interpretation, the principles of quantum theory could now finally be expressed. Here is a (very simplified) summary of the basics of quantum mechanics:

1. Start with the wave function $\Psi(x)$, which describes an electron located at the point x.
2. Insert the wave into the Schrödinger equation $H\Psi(x) = i(h/2\pi) \partial_t \Psi(x)$. (H, which is known as the Hamiltonian, corresponds to the energy of the system.)
3. Each solution of this equation is labeled an index n, so in general, $\Psi(x)$ is a sum or superposition of all these multiple states.
4. When a measurement is made, the wave function "collapses," leaving only one state $\Psi n(x)$, i.e., all the other waves are set to zero. The probability of finding the electron in this state is given by the absolute value of $\Psi n(x)$.

These simple rules can in principle derive everything known about chemistry and biology. What is controversial about quantum mechanics is contained in the third and fourth statements. The third statement says that in the subatomic world, an electron can exist simultaneously as the sum of different states, which is impossible in Newtonian mechanics. Before a measurement is made, in fact, the electron exists in this netherworld as a collection of different states.

But the most crucial and outrageous statement is number four, which holds that only after a measurement is made will the wave finally "collapse" and yield the correct answer, giving the probability of finding the electron in that state. One cannot know which state the electron is in until a measurement is made.

This is called the measurement problem.

To refute the last statement, Einstein would say, "God does not play dice with the universe." But according to legend, Niels Bohr fired back, "Stop telling God what to do."

It is precisely postulates 3 and 4 that make quantum computers possible. The electron is now described as the simultaneous sum over different quantum states, which gives quantum computers their calculational power. While classical computers only sum over just 0s and 1s, quantum computers sum over all quantum states $\Psi_n(x)$ between 0 and 1, which vastly increases the number of states and therefore their range and power.

Ironically, Schrödinger, whose equations started the whole quantum mechanics bandwagon in the first place, began to denounce this version of his own theory. He regretted the fact that he had anything to do with it. He thought that a simple paradox that demonstrated the absurdity of this radical interpretation would destroy it forever, and it started with a cat.

Schrödinger's Cat

Schrödinger's cat is the most famous animal in all of physics. Schrödinger believed it would demolish this heresy once and for all. Imagine, he wrote, there is a cat in a sealed box, which contains a vial of poison gas. This vial is connected to a hammer, which is attached to a Geiger counter next to a quantity of uranium. If an atom of the uranium decays, it activates the Geiger counter, which sets off the hammer, thus releasing the poison and killing the cat.

Now here is the question that has baffled the world's top physicists for the past century: *Before you open the box, is the cat dead or alive?*

A Newtonian would say that the answer is obvious: common sense says that the cat is either dead or alive, but not both. You can only be in one state at a time. Even before you opened the box, the cat's fate was already predetermined.

However, Werner Heisenberg and Niels Bohr had a radically different interpretation.

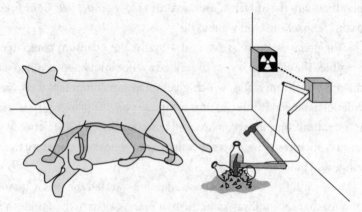

Figure 4: Schrödinger's cat
In quantum mechanics, to describe a cat in a sealed box containing a vial of poison gas, and a hammer triggered by a Geiger counter, one must add the wave function of a dead cat with that of a live cat. Before you open the box, the cat is neither dead nor alive. The cat is in a superposition of two states. Even today physicists debate the question of how the cat can be dead and alive at the same time.

They said that the cat is best represented by the sum of two waves: the wave of the live cat and the dead cat. When the box is still sealed, the cat can only exist as the superposition or sum of two waves simultaneously representing a dead and a live cat.

But is the cat dead or alive? As long as the box is sealed, this question makes no sense. *In the microworld, things do not exist in definite states, but only as the sum of all possible states.* Finally, when the box is opened and you observe the cat, the wave miraculously collapses and reveals the cat as being either dead or alive, but not both. So the process of measurement connects the microworld and the macroworld.

This has deep philosophical implications. Scientists spent many centuries arguing against something called solipsism, the idea that philosophers like George Berkeley believed that objects do not really exist unless you observe them. The philosophy can be summarized as "To be is to be perceived." If a tree fell in the forest but no one was there to listen to it fall, then perhaps the tree never fell at all. Reality, in this picture, is a human construct. Or, as the poet John Keats once said, "Nothing ever becomes real till it is experienced."

The quantum theory, however, made this situation worse. In quantum theory, before you look at a tree, it can exist in all possible states, such as firewood, lumber, ash, toothpicks, a house, or sawdust. However, when you actually look at the tree, all the waves representing these states miraculously collapse into one object, the ordinary tree.

But since an observer requires consciousness, this means that, in some sense, consciousness determines existence. The followers of Newton were aghast that solipsism was creeping back into physics.

Einstein hated this idea. Like Newton, Einstein believed in "objective reality," which means that objects exist in definite, well-defined states, i.e., you cannot be in two places at the same time. This is also known as Newtonian determinism, the idea that, as we saw earlier, you can precisely determine the future using fundamental physical laws.

Einstein would often make fun of the quantum theory. Whenever guests would visit his house, he would ask them to look at the moon. Does the moon exist, he would ask, because a mouse looks at it?

Microworld Versus Macroworld

Mathematician John von Neumann, who helped develop the physics of the quantum theory, believed that there was an invisible "wall" that separated the microworld from the macroworld. They each obeyed different laws of physics, but you could prove that you could freely move the wall, back and forth, and the outcome of any experiment remained the same. In other words, the microworld and the macroworld obeyed two different sets of physics, but this did not affect measurements because it did not matter precisely where you chose to separate the micro- and macroworlds.

When asked to clarify the meaning of this wall, he would say, "You just get used to it."

But no matter how crazy the quantum theory appeared to be, its experimental success was indisputable. Many of its predictions (when predicting the properties of electrons and photons in what is called quantum electrodynamics) fit the data to within one part in 10 billion, making it the most successful theory of all time. The atom, once the most mysterious object in the universe, was suddenly spilling out its deepest secrets. The next generation of physicists who embraced the quantum theory were awarded scores of Nobel Prizes. Not a single experiment violated the quantum theory.

The universe was undeniably a quantum universe.

But Einstein summed up the successes of the quantum theory by stating, "The more successful the quantum theory becomes, the sillier it looks."

What the critics of quantum mechanics most objected to was this artificial separation between the macroworld that we live in and the weird, preposterous world of the quantum. The critics said that there must be a smooth continuum from the microworld to the macroworld. In reality, there is no "wall."

For example, if we can hypothetically live in a completely quantum world, it means that everything we know about common sense is wrong. For example:

- We can be two places at the same time.
- We can disappear and reappear somewhere else.
- We can walk through walls and penetrate barriers effortlessly, which is called tunneling.

- People who have died in our universe might be alive in another.
- When we walk across a room, we actually simultaneously take all the infinite number of possible paths across the room, no matter how bizarre.

As Bohr would say, "Anyone who is not shocked by the quantum theory does not understand it."

All of this is grist for *The Twilight Zone*. But miraculously, this is precisely what electrons do, except they do it mainly inside the atom, where we can't see them doing these gymnastics. That is why we have lasers, transistors, digital computers, the internet. Isaac Newton would be shocked if he could somehow see all the atomic gyrations that electrons perform to make computers and the internet possible. But the modern world would collapse if we outlawed the quantum theory and set Planck's constant to zero. All the miraculous electronic devices in your living room are possible precisely because electrons can perform these fantastic tricks.

But we never see these effects in our life, because we consist of trillions upon trillions of atoms, where these quantum effects average each other out, and because the size of these quantum fluctuations is Planck's constant h, which is a very small number.

Entanglement

In 1930, Einstein had had enough. At the Sixth Solvay Conference in Brussels, Einstein decided he would go head-to-head and challenge Niels Bohr, the leading proponent of quantum mechanics. It was to be the Clash of Titans, with the greatest physicists of the age debating the very destiny of physics and the nature of reality. What was at stake was the very meaning of existence. Physicist Paul Ehrenfest would write, "I will never forget the sight of the two opponents leaving the university club. Einstein, a majestic figure, walking calmly with a faint ironical smile, and Bohr trotting along by his side, extremely upset." Later, Bohr was so shaken that he could be seen muttering to himself, "Einstein . . . Einstein . . . Einstein . . ."

Physicist John Archibald Wheeler recalled, "It was the greatest debate in intellectual history that I know about. In thirty years, I never heard of a debate between two greater men over a longer period of time on a deeper issue with deeper consequences for understanding this strange world of ours."

Time and time again, Einstein would bombard Bohr with the paradoxes of the quantum theory. It was merciless. Bohr would be temporarily dazed by each barrage of criticisms, but the next day he would collect his thoughts and give a cogent, airtight response. Once, Einstein caught Bohr in another paradox concerning light and gravity. It seemed as if Bohr had finally been checkmated. But ironically, Bohr was able to find the flaw in Einstein's reasoning by quoting from Einstein's own theory of gravity.

The verdict from most physicists was that Bohr had successfully refuted every argument made by Einstein at the famous Solvay Conference. But Einstein, perhaps smarting from this setback, would try once more to topple the quantum theory.

Five years later, Einstein mounted his final counterattack. With his students Boris Podolsky and Nathan Rosen, they made one last valiant attempt to smash the quantum theory once and for all. The EPR paper, named after its authors, was to be the final blow against the quantum theory.

One unforeseen by-product of this fateful challenge would be the quantum computer.

Imagine, they said, two electrons that are coherent with each other, meaning they are vibrating in unison, i.e., with the same frequency but shifted by a constant phase. It is well known that electrons have spin (which is the reason why we have magnets). If we have two electrons with a total spin of zero, and if we let one electron spin, say, clockwise, then the other electron spins counterclockwise because the net spin is zero.

Now separate the two electrons. The sum of the spins of the two electrons must still be zero, even if one electron is now on the other side of the galaxy. But you cannot know how it is spinning before you take a measurement. But strangely, if you measure the spin of one electron and find it is spinning clockwise, then you instantly know that its partner on the other side of the galaxy must be spinning counterclockwise. This information traveled instantaneously between the two electrons, faster than the speed of light. In other words, as you separate these two electrons, an invisible umbilical cord emerges between them, allowing communication to travel through the cord faster than the speed of light.

But, claimed Einstein, since nothing can go faster than the speed of light, this was in violation of special relativity, and hence quan-

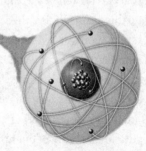

Figure 5: Entanglement
When two atoms are next to each other, they can vibrate coherently, in unison, with the same frequency but shifted by a constant phase. But if we separate them and jiggle one of them, they are still coherent, and information of the disturbance can travel between them faster than the speed of light. (But this does not violate relativity, since the information that breaks the light barrier is random.) This is one reason why quantum computers are so powerful, because they compute simultaneously over all these mixed states.

tum mechanics is incorrect. This was the killer argument that disproved the quantum theory, Einstein thought. He rested his case. The "spooky action at a distance" created by entanglement was just an illusion, he claimed.

Einstein thought he delivered the coup de grâce that would quash the quantum theory once and for all. But in spite of all the experimental successes of the quantum theory, the so-called EPR paradox was unresolved for decades because it was too difficult to perform in the laboratory. But over the years, this experiment was finally performed in several ways, in 1949, 1975, and 1980, and each time the quantum theory came out correct.

(But does this mean that information can travel faster than light, violating special relativity? Einstein has the last laugh here. No, although the information that traveled between the two electrons was instantaneously transmitted, it was also random information, and hence useless. This means you cannot send useful codes containing a message faster than light using the EPR experiment. If you actually analyze the EPR signal, you only find gibberish. So information can travel instantly between coherent particles, but useful information that carries a message cannot go faster than light.)

Today, this principle is called entanglement, the idea that when two objects are coherent with each other (vibrating in the same way), then they remain coherent, even if separated by vast distances.

This has major implications for quantum computers. It means that, even if the qubits in a quantum computer are separated, they can still interact with each other, which is responsible for the fantastic computational ability of quantum computers.

This gets at the essence of why quantum computers are so unique and useful. An ordinary digital computer, in a sense, is like several accountants toiling away independently in an office, each doing one calculation separately, and handing off their answers from one to another. But a quantum computer is like a roomful of interacting accountants, each one simultaneously computing, and, importantly, communicating with each other via entanglement. So we say that they are coherently solving this problem together.

Tragedy of War

Unfortunately, this vibrant intellectual debate was interrupted by the rising tide of world war. Suddenly, the scholarly discussions about the quantum theory became deadly serious, as both Nazi Germany and the U.S. instituted crash programs to develop the atomic bomb. The Second World War would have devastating consequences for the physics community.

Planck, witnessing the wholesale migration of Jewish physicists from Germany, personally met with Adolf Hitler, pleading him to stop the persecution of Jewish physicists, which was destroying German physics. However, Hitler became enraged at Planck and screamed at him.

Afterward, Planck said, "You cannot reason with such a man." Sadly, one of Planck's sons, Erwin, was later involved with a plot to assassinate Hitler. He was caught and tortured. Planck tried to save his son's life by appealing directly to Hitler. But Erwin was executed in 1945.

The Nazis put a price on Einstein's head. His picture was featured on the cover of a Nazi magazine, with the caption "Not Yet Hanged." He fled Germany in 1933, never to return.

Erwin Schrödinger, who witnessed a Jewish man being beaten by the Nazis in the streets in Berlin, tried to stop the attack, only to be beaten himself by the SS. Shaken, he left Germany and accepted a position at Oxford University. But he stirred controversy there because he came with his wife and his mistress. He then was offered a position at Princeton, but historians conjecture he turned it down because of his unorthodox living arrangements. He eventually ended up in Ireland.

Niels Bohr, one of the founders of quantum mechanics, had to flee for his life to the United States and almost died in the process of escaping Europe.

Werner Heisenberg, perhaps the greatest quantum physicist in Germany, was put in charge of developing the atomic bomb for the Nazis. However, his laboratory had to move repeatedly as it was bombed by the Allies. After the war, he was arrested by the Allies. (Fortunately, Heisenberg did not know one key number, the probability for splitting the uranium atom, so he had difficulty building the atomic bomb and the Nazis never developed a nuclear weapon.)

In the tragic aftermath of war, people began to realize the enormous power of the quantum, which was unleashed in the skies over Hiroshima and Nagasaki. Suddenly, quantum mechanics was not just the plaything of physicists, but something that could unlock the secrets of the universe and hold the destiny of the human race.

But out of the ashes of war, a new quantum invention was on the horizon that would alter the very fabric of modern civilization: the transistor. Perhaps the enormous power of the atom could be used to bring peace.

CHAPTER 4

DAWN OF QUANTUM COMPUTERS

The transistor is a paradox.

Usually, the larger an invention, the more powerful it is. Huge double-decker jetliners can carry loads of passengers halfway around the world in a matter of hours. Rockets today are towering inventions able to send multiton payloads to Mars. The nearly seventeen-mile-long Large Hadron Collider cost over $10 billion and may one day unravel the secret of the Big Bang. Its circumference is so big that much of the city of Geneva can be put inside the perimeter of the machine.

Yet the transistor, perhaps the most important invention of the twentieth century, is so small that billions of them can fit on your fingernail. It is not an exaggeration to say that it has revolutionized every aspect of human society.

So sometimes smaller is better. For example, sitting on your shoulders is the most complex object in the known universe, the human brain. Consisting of 80 billion neurons, each connected to about 10,000 other neurons, the human brain in its complexity exceeds anything known to science.

So both a microchip made of billions of transistors and the human brain can be held in your hand, yet they are the most sophisticated objects that we know about.

Why is that? Their incredibly small size hides the fact that you can store and manipulate vast amounts of information within them. Furthermore, the way that this information is stored resembles a Turing machine, giving them tremendous calculational power. A microchip is the heart of a digital computer with a finite input tape (though Turing machines in principle can have an infinite tape). And the brain is a learning machine or neural network that constantly modifies itself as it learns new things. A Turing machine can be modified so it too can learn like a neural network.

But if the power of the transistor comes from being microscopic, then the next question is: How small can you make a computer? What is the smallest transistor?

Birth of the Transistor

Three physicists won the Nobel Prize in 1956 for the creation of this wonder device: Bell Laboratories scientists John Bardeen, Walter Brattain, and William Shockley. Today, a replica of the world's first transistor is on display in a glass case in the Smithsonian Museum in Washington. It is a crude, awkward-looking device, but delegations of scientists from around the world are known to approach this transistor with silent reverence, and some even bow in front of it, as if it were some deity. Bardeen, Brattain, and Shockley used a new quantum form of matter, called the semiconductor. (Metals are conductors, which allow for the free flow of electrons. Insulators, like glass, plastic, or rubber, do not conduct electricity. Semiconductors are in between and can both carry and stop the flow of electrons.)

The transistor exploits this crucial property. It is the successor to the old vacuum tube that was ingeniously used by Turing and others. As we have seen, both a vacuum tube and transistor can be roughly compared to a valve controlling the flow of water in a pipe. With a small valve, you can control the much larger flow of water going through a pipe. You can either shut it off, which corresponds to zero, or leave it open, which corresponds to one. In this way, you can precisely control the flow of water in a complex series of pipes. If you now replace the valve with a transistor and the water pipes with wires conducting electricity, you can create a digital transistorized computer.

While a transistor resembles a vacuum tube in this way, that is where the similarities stop. Vacuum tubes are famous for being crude and temperamental. (As a child, I remember having to take apart my old TV set, remove all the tubes by hand, and tediously test each one at the supermarket to see which one blew out.) They were bulky, unreliable, and rapidly wore out.

A transistor on the other hand, made of thin, silicon wafers, can be sturdy, cheap, and microscopic in size. They can be mass-produced in the same way that a T-shirt is made today.

T-shirts are usually made from a plastic template, which has the image you want cut out in the plastic. The template is placed over the T-shirt, and then a spray can sprays paint over the template. When you remove the template, the image is transferred onto the T-shirt.

A transistor is created similarly. First, you start with a template where the image of the circuits you want has been carved out. Then you place the template over the silicon wafer. You then apply a beam of ultraviolet radiation onto the template, so the image in the template is transferred onto the silicon wafer. You then remove the template and add acid. The silicon chip is specially treated chemically so that, when you apply the acid, it burns the image you desire in the wafer.

The advantage is that these images can be as small as the wavelength of ultraviolet light, which is a bit larger than an atom. This means that a typical chip used in a computer can have a billion transistors. Today, producing transistors is big business, affecting the economies of entire nations. The most advanced factories for producing transistors cost several billion dollars each.

In some sense, a microchip can be compared to the roadways in a large city. The constant flow of cars is like electrons traveling along the etched circuits. The traffic signals that regulate the flow of traffic correspond to transistors. A red light stopping traffic is like 0, while a green light that allows traffic to flow is like a 1.

When we etch more and more transistors onto a chip, it is as if we are shrinking the size of each city block to increase the number of cars and traffic lights. But there is a limit to how tightly you can pack the roads in a given area. Eventually, the city block becomes so tiny that cars spill out onto the sidewalk. This equates to short circuits if the layers of silicon become too thin.

As the width of the components of a silicon chip approaches the size of an atom, the Heisenberg uncertainty principle kicks in, and the electrons' positions become uncertain, causing them to leak out and short the circuit. Further, the heat generated by so many transistors packed in one place is sufficient to cause it to melt.

In other words, all things must pass, including the Silicon Age. A new age may be dawning: the Quantum Age.

And one of the most famous physicists of the twentieth century paved the way.

Genius in Action

Richard Feynman was one of a kind. There will probably never be another physicist like him.

On one hand, Feynman was a charismatic showman, fond of amusing audiences with outrageous stories of his past and his crazy antics. In his rough accent, he sounded like a truck driver as he told colorful tales about his life.

He prided himself on being an expert in picking locks and cracking safes, even successfully unlocking the safe containing the secret of the atomic bomb while working at Los Alamos (and setting off a huge alarm in the process). Always interested in new, quirky experiences, he once sealed himself in a hyperbaric chamber to find out if he could leave his body and see himself floating from a distance. And he would love to play his bongos at all hours of the day.

Listening to him, one might almost forget that he won the Nobel Prize in Physics in 1965 and was probably one of the greatest physicists of his generation, laying the complex groundwork for a relativistic theory of electrons interacting with photons. This theory, called quantum electrodynamics or QED, is accurate to one part in 10 billion; so of all the various quantum measurements that have been made, it is the most successful. Other physicists would intently listen to his every word, hoping to absorb the insights that might also win them fame and glory.

Birth of Nanotech

Above all, Feynman was a visionary.

Feynman realized that computers were becoming smaller and smaller. So he asked himself a simple question: How small can you make a computer?

He realized that in the future, transistors would become so small they would eventually become the size of atoms. In fact, he conjectured, the next frontier for physics could be to create machines as small as atoms, pioneering a growing field now called nanotechnology.

What limit does quantum mechanics place on tweezers, hammers, and wrenches that are the size of atoms? What is the ultimate limitation to a computer that computes on transistors the size of atoms?

He realized that in the atomic realm, new fantastic inventions are possible. The current laws of physics that we use on the macroscale become obsolete at the atomic scale, and we have to open our minds to entirely new possibilities. His ideas were first expressed in a speech he gave to the American Physical Society at Caltech in 1959, titled "There's Plenty of Room at the Bottom," anticipating the birth of a new science.

In that pioneering article, he asked, "Why cannot we write the entire 24 volumes of the Encyclopedia on the head of a pin?"

His basic idea was simple: to create tiny machines that could "arrange the atoms the way we want." Any tool that we use in our workshop would be miniaturized to the size of fundamental particles. Mother Nature manipulates atoms all the time. Why can't we?

He summarized his idea for quantum computers by saying, "Nature isn't classical, dammit, and if you want to make a simulation of nature, you'd better make it quantum mechanical."

This is a profound observation. Classical digital computers, no matter how powerful, can never successfully simulate a quantum process. (Bob Sutor, vice president of IBM, likes to make this comparison: for a classical computer to re-create a one-to-one simulation of a simple molecule, like caffeine, it would require 10^{48} bits of information. That huge number is 10 percent of the number of atoms making up the planet earth. Hence, classical computers cannot successfully simulate even simple molecules.)

In his article, Feynman introduced a number of amazing ideas. He proposed a robot so small it could float in your bloodstream and tend to medical problems. Feynman called this "swallowing the doctor." It would function like a white blood cell, roaming the body looking for bacteria and viruses that it could zap. It would also perform surgery while circulating in your body. So medicine would be practiced from inside the body, not outside. You would never have to cut the skin or worry about pain and infections because surgery would be done from within.

He was prophetic in his vision, even claiming that one day it would be possible to invent a super microscope to "see" atoms. (This actually was invented later in 1981, a few decades after he made that prediction, in the form of scanning tunneling microscopes.)

His vision was so fantastic that his speech was largely ignored for decades afterward. This was a shame, because he was far ahead of his time. Yet today, many of his predictions have come to pass.

He even offered a prize of $1,000 to anyone who could invent one of two things: The first challenge was to miniaturize a page of a book so that only an electron microscope could see it. The second $1,000 prize was to create an electric motor that could fit into a $1/64$-inch cube. (Two inventors later claimed both prizes, although they did not fulfill the precise requirements of the contest.)

Another of his predictions has been made possible with the discovery of nanomaterials such as graphene, consisting of a sheet of carbon that is just one atom thick. Graphene was discovered by two Russian scientists working in Manchester, England, Andre Geim and Konstantin Novoselov, who noticed that Scotch tape could peel off a thin layer of graphite. By doing this peeling process repeatedly, they found that eventually they could peel off a single layer of carbon one atom thick. For this simple but remarkable breakthrough, they won the Nobel Prize in 2010. Because the carbon atoms are so tightly packed in a symmetrical array, it is the strongest substance known to science, stronger than diamonds. A sheet of graphene is so strong that if you balanced an elephant on the end of a pencil, and balanced the pencil on a sheet of graphene, it would not tear.

Small quantities of graphene are easy to manufacture, but making large-scale amounts of pure graphene is extremely difficult. In principle, though, pure graphene is strong enough to build a skyscraper or a bridge so thin it would be invisible. A long fiber of graphene might be so strong that it could support a space elevator which takes you into space with the push of a button, like an elevator to heaven. (The space elevator would be suspended on a cable of graphene, which, like twirling a ball on a string, never falls down because it rotates around the earth due to the spin of the planet.) Also, graphene can conduct electricity. In fact, some of the world's smallest transistors can be made from tiny amounts of graphene.

Feynman also realized the enormous advances that could be possible with a quantum computer, which would have enormous computational power. Earlier, we saw that if you add just one more qubit to a quantum computer, its power doubles. So a quantum computer made of 300 atoms would have 2^{300} the power of a quantum computer with one qubit.

Feynman's Path Integral

Yet another accomplishment of Feynman would change the course of physics. He would find a startling new way to reformulate the entire theory of quantum mechanics.

It all started when he was in high school. He loved to calculate things and solve puzzles. One of his trademarks was rapidly calculating the answer to a puzzle in several different ways. If he got stuck in one direction, he knew mathematical tricks to solve the problem another way. He was famous for saying that the goal of every physicist "is to prove yourself wrong as soon as possible." In other words, swallow your pride and admit that what you are doing may be a dead end, and prove it as soon as possible so you can move on to the next idea.

(As a research physicist, I actually think about this statement often. At some point physicists may have to admit that maybe their pet idea is wrong, and that they should quickly try a new approach.)

Because the young Feynman was always ahead of his class in science, his high school teacher thought of clever ways to keep him amused so he wouldn't get bored. The teacher would challenge him with curious but profound lessons from physics.

One day his teacher introduced him to something called the principle of least action, which allows for a radical reinterpretation of all of classical physics. The teacher noted that if a ball rolls down a hill, there are an infinite number of ways it can possibly roll, but there is only one path that it actually takes. How does it know which path to take?

Three hundred years earlier, Newton solved this question. He would say: calculate the forces acting on the ball at one instant, and then use his equations to determine where it will go in the next instant. Then repeat the process. By sewing together all these successive instants of time, microsecond after microsecond, one can trace out its entire path. Even today, three hundred years later, this is how physicists predict the motion of stars, planets, rockets, cannon balls, and baseballs. This is the fundamental basis of Newtonian physics. Almost all of classical physics is done this way. And the mathematics of adding up all these incremental motions is called calculus, also invented by Newton.

But then the teacher introduced a bizarre way to look at this. He said, draw *all* possible paths the ball may take, no matter how strange. Some of these paths may be absurd, like taking a trip to the moon or Mars. Some paths may even go to the ends of the universe. For each path, calculate what is called the action. (The action is similar to the energy of the system. It is the kinetic energy minus the potential energy.) Then the path of the ball will be the one with the smallest value for the action. In other words, somehow the ball "sniffs" out all possible paths, even crazy ones, and then "decides" to take the path with the least action.

When you do the math, you get precisely the same answer as Newton. Feynman was amazed. In this simple demonstration, one could summarize all of Newtonian physics without complicated differential equations—all you had to do was find the path with the least action. This delighted Feynman because now he had two equivalent ways to solve all of classical mechanics.

In other words, in the old Newtonian picture, the path of a ball is only determined by the forces acting on the ball at that precise point in space and time. Distant points do not affect the ball at all. But in this new picture, the ball is suddenly "aware" of all possible paths the ball may take, and "decides" to take the one with the least action. How could the ball "know" how to analyze these billions of paths and select just the right one?

(For example, why does a ball fall to the floor? Newton would say that there is a gravitational force pushing the ball to the ground, microsecond by microsecond. Another explanation is to say that the ball somehow sniffs out all possible paths, and then decides to take the path of least action or energy, which is straight down.)

Years later, when Feynman was doing his Nobel Prize–winning work, he would return to this approach from high school. The principle of least action worked for classical, Newtonian physics. Why not generalize this strange result to the quantum theory?

Quantum Sum Over Paths

He realized that in a quantum computer this would have tremendous calculational power. Think of a maze. If a classical mouse were put in a maze, it would tediously try out many possible paths, one after the other, in sequence, which is extremely slow. But if you put a quantum mouse in a maze, it simultaneously sniffs out all possible paths at the same time. When applied to a quantum computer, this principle exponentially increases its power.

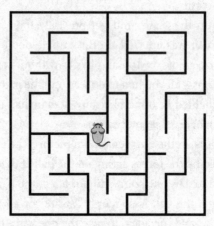

Figure 6: Sum over paths
A mouse in a classical maze has to decide which way to turn at every juncture, one decision at a time. But a quantum mouse in a maze can, in some sense, analyze all possible paths simultaneously. This is one reason why quantum computers are exponentially more powerful than ordinary classical computers.

So Feynman rewrote the quantum theory in terms of the principle of least action. In this view, subatomic particles "sniff out" all possible paths. On each path he put a factor related to the action and Planck's constant. Then he summed or integrated over all possible paths. This is now called the path integral approach, because you are adding up contributions from all the paths an object can take.

Much to his shock, he found he could derive the Schrödinger equation. In fact, he found that he could summarize *all* of quantum physics in terms of this simple principle. So decades after Schrödinger introduced his wave equation by magic, with no derivation, Feynman was able to unify the entirety of quantum mechanics, including the Schrödinger equation, using this path integral approach.

Usually, when I teach quantum mechanics to PhD students in physics, I start by presenting the Schrödinger equation as if it simply came out of nowhere, like from a magician's hat. When students ask me where did this equation come from, I simply shrug my shoulders and say that the equation just is. But later in the course, when we finally discuss path integrals, I explain to the students that all of quantum theory can be reformulated using Feynman path integrals, by summing the action over all possible paths, no matter how crazy they are.

Not only do I use Feynman path integrals in my professional work, I also think about them at home sometimes as I walk across a room. As I move over a carpet, I have a strange, eerie feeling knowing that many copies of myself are also walking across the same carpet, and each one thinks that they are the only person walking across the room. Some of these copies even went to Mars and back.

As a physicist, I work with relativistic versions of Schrödinger's equation, which is called quantum field theory, i.e., the quantum theory of subatomic particles at high energies. The very first thing I do when calculating with quantum field theory is to follow Feynman and start with the action. I then calculate over all possible paths to get the equations of motion. So Feynman's path integral approach, in some sense, has swallowed up all of quantum field theory.

But this formalism is not just a trick; it also has some profound implications for life on earth. We saw earlier that quantum computers have to be kept at near absolute zero. But Mother Nature can perform marvelous quantum reactions at room temperature (such as photosynthesis and the fixing of nitrogen for fertilizers). Under classical physics, there is so much noise and jostling of atoms at room temperature that many chemical processes should be impossible in those conditions. In other words, photosynthesis violates the laws of Newton.

So how does Mother Nature solve the problem of decoherence, the most difficult problem in quantum computers, to enable photosynthesis at room temperature?

By summing over all paths. As Feynman showed, electrons can "sniff" out all possible paths to do their miraculous work. In other words, photosynthesis, and hence life itself, may be a by-product of Feynman's path integral approach.

Quantum Turing Machine

In 1981, Feynman emphasized that only a quantum computer can truly simulate a quantum process. But Feynman did not elaborate on precisely how a quantum computer might be built. The next person who picked up the torch was David Deutsch of Oxford University. Among other achievements, he was able to answer the question: Can you apply quantum mechanics to a Turing machine? Feynman had hinted at this problem, but never wrote down the equations for a quantum Turing machine. Deutsch went on to fill in all the details. He even designed an algorithm that could run on this hypothetical quantum Turing machine.

A Turing machine, as we have seen, is a simple classical device, based on a processor, that turns the number on an infinitely long tape into another number and therefore performs a series of mathematical operations. The beauty of a Turing machine is that it summarizes all the properties of a digital computer in a simple, compact form, which can then be rigorously studied by mathematicians. The next step is to add the quantum theory to Turing's invention, which would allow scientists to study the bizarre properties of quantum computers in a rigorous way. In a quantum Turing machine, thought Deutsch, one replaces a classical bit with a quantum qubit. This introduces several important changes.

First, the basic manipulations of the Turing machine (e.g., replacing a 0 with a 1 and vice versa, and moving the tape forward or backward) remain roughly the same. But the bits are radically altered. They are no longer 0 or 1. In fact, they can use the bizarre quantum property of superposition (the ability to be in two distinct states at the same time) to create a qubit, which can assume values between 0 and 1. And because all the qubits in a quantum Turing machine are entangled, what happens to one qubit can influence other qubits that are far away. Lastly, to get a number at the end of the calculation, one has to "collapse the wave," so that qubits give us a collection of 0s or 1s back again. In this way, we can extract real numbers and answers from the quantum computer.

In the same way that Turing was able to make the field of digital computers rigorous by introducing the precise rules of Turing machines, Deutsch helped make the foundation of quantum computers rigorous. By isolating the essence of how qubits are manipulated, he helped standardize work on quantum computers.

Parallel Universes

But not only is Deutsch well known for developing the concept of quantum computers, he also takes seriously the deep philosophical questions raised by them. In the usual Copenhagen interpretation of quantum mechanics, one has to make an observation to finally determine where an electron is. Before an observation is made, the electron is in a fuzzy mixture of several states. But when the state of the electron is measured, the wave function magically "collapses" down to one physical state. This is how one extracts numerical answers from a quantum computer.

But this "collapsing" has haunted quantum physicists for the past century. This process of "collapsing" the wave seems so alien, so contrived and artificial, yet it is the crucial process that allows one to leave the quantum world and enter our macroscopic world. Why does it snap to attention just when we decide to look at it? It is the bridge between the micro- and macroworlds, but it is a bridge with huge philosophical holes in it.

Still, it works. No one can deny this.

But many scientists feel uneasy knowing that all our knowledge of the world is built on this uncertain foundation, like shifting sand that might one day blow away. Over the past decades, numerous proposals have been made to clarify this problem.

Perhaps the most outrageous of these proposals was made in 1956 by graduate student Hugh Everett. We recall that the quantum theory can be summarized in roughly four broad principles. The last one is the sticking point, where we "collapse" the wave function to decide what state the system is in. Everett's proposal was daring and controversial: his theory says simply to *drop the last statement that says the wave "collapses" so it never does at all.* Each possible solution continues to exist in its own reality, producing, as the theory is known, "many worlds."

Like a river branching into many smaller tributaries, the various waves of the electron keep propagating merrily along, splitting and resplitting again and again, branching off into other universes forever. In other words, there are an infinite number of parallel universes, none of which ever collapses. Each branch of this multiverse appears real as any other, but they represent *all* possible quantum states.

The microcosm and the macrocosm therefore obey the same equations, since there is no collapse and no "wall" separating them anymore.

For example, think of an ocean wave. Inside, it actually consists of thousands of smaller waves. The Copenhagen interpretation means selecting just one of these smaller waves and throwing away the rest. But the Everett interpretation says to let all the waves exist. Then the waves will continue to branch off smaller waves, which in turn branch off even more waves.

This idea is very convenient. You never have to worry about waves "collapsing," because they never do. So this formulation is simpler than the standard Copenhagen interpretation. It is neat, elegant, and remarkably simple.

Many Worlds

However, Everett's and Deutsch's theories challenge the very nature of reality. The many worlds theory is one that overturns our conception of existence itself. Its consequences are staggering.

For example, think of all the times that you had to make a crucial decision in your life, such as what job to apply for, whom to marry, whether to have children. One may spend hours on a lazy afternoon thinking of all the things that might have been. The many worlds theory says that there is a parallel universe with a copy of yourself, living out a totally different life story. In one universe, you may be a billionaire contemplating your next headline-making adventure. In another universe, you may be a pauper wondering where your next meal will come from. Or you may be living somewhere in between, toiling at a boring, tedious job with a low, steady income but no future. In each universe, you insist that your universe is the real one, and all the others are fake. Now imagine this at a quantum level. Each individual atomic action splits our universe into multiples of itself.

In Robert Frost's poem "The Road Not Taken," he wrote about something that everyone has thought about in their daydreams. We wonder what might have happened during times in our lives when we made a critical choice. These momentous decisions may affect our lives ever after. He wrote:

> Two roads diverged in a yellow wood,
> And sorry I could not travel both
> And be one traveler, long I stood
> And looked down one as far as I could
> To where it bent in the undergrowth.

He ends the poem by concluding that his decision had epic consequences for his life, that the road less traveled was a turning point. He concludes:

> I shall be telling this with a sigh
> Somewhere ages and ages hence:
> Two roads diverged in a wood, and I—
> I took the one less traveled by,
> And that has made all the difference.

This extends not just to your life, but to the entire world. In the TV series *The Man in the High Castle,* based on the novel by Philip K. Dick, the universe has split in half. In one universe, an assassin tried to kill Franklin D. Roosevelt, but his gun jammed, and FDR lived on to lead the Allies to victory during World War II. But in another universe, the gun did not misfire, and the president was killed. A weak vice president takes over, and the U.S. is defeated. The Nazis then occupy the East Coast of the U.S., while the Japanese Imperial Army seizes the West Coast.

What separates these very different, divergent universes is the jamming of a single bullet. But bullets can misfire because of tiny imperfections in their chemical propellant, perhaps caused by quantum defects in the molecular structure of the explosive. So one quantum event may separate these two universes.

Unfortunately, Everett's idea was so radical, so out of this world, that it was uniformly ignored by physicists for decades. Only recently has it experienced a revival as physicists rediscover his work.

Everett's Many Worlds

Hugh Everett III was born in 1930 to a military family. His father, who helped raise him after his parents divorced, was a lieutenant colonel on the general staff during World War II. After the war, his father was stationed in West Germany, where Hugh joined him.

At an early age, he showed an interest in physics. He even wrote a letter to Einstein, who actually replied to his question about a long-standing philosophical problem as follows:

Dear Hugh,

There is no such thing like an irresistible force and immovable body. But there seems to be a very stubborn boy who has forced his way victoriously through strange difficulties created by himself for this purpose.

Sincerely yours,
A. Einstein.

At Princeton, he finally pursued his scientific interests, which centered on two areas. First, in how science could affect military affairs, e.g., using game theory to understand warfare. And second was trying to understand the paradoxes of quantum mechanics. His PhD advisor was John Archibald Wheeler, the same thesis advisor who mentored Richard Feynman. Wheeler was one of the grand old men of physics and had worked with Bohr and Einstein.

Everett was dissatisfied with the traditional Copenhagen interpretation of quantum mechanics, where the wave function mysteriously "collapses" and determines the state of the macroworld that we live in.

His solution was radical, yet also simple and elegant. Wheeler immediately grasped the importance of his student's work, but he was also a realist. He knew that this theory would get soundly trashed by the establishment. So on several occasions Wheeler would ask Everett to tone down the theory so that it didn't seem so outrageous. Everett did not like this at all, but, because he was just a graduate student, he went along with these revisions. Wheeler would sometimes try to discuss his student's theory with other prominent physicists, but usually got a cold reception.

In 1959, Wheeler even arranged to have Everett meet Niels Bohr himself in Copenhagen. It was Wheeler's last attempt to gain some recognition for his student's work. But he was like a lamb entering the lion's den. The meeting was a disaster. Belgian physicist Léon Rosenfeld, who was there, said that Everett was "undescribably [*sic*] stupid and could not understand the simplest things in quantum mechanics."

Everett would later recall that this meeting was "hell . . . doomed from the beginning." Even Wheeler, who had tried to give Everett's theory a fair hearing among top physicists, eventually abandoned the theory, saying it had "too much baggage."

With all the biggest names in physics allied against him, a job in theoretical physics was highly unlikely, so he went back to his military studies and got a job with the Pentagon's Weapons Systems Evaluation Group. From there, he would do top secret research on Minuteman missiles, nuclear war and fallout, and the military applications of game theory.

Rebirth of Parallel Universes

Meanwhile, during the years he was working on nuclear warfare, his ideas began to slowly percolate in the physics community. One problem arose when physicists tried to apply quantum mechanics to the entire universe, i.e., to create a quantum theory of gravity.

In quantum mechanics, we start with a wave that describes how an electron can be in many parallel states at the same time. At the end, the observer makes a measurement from the outside and collapses the wave function. But we encounter problems when applying this process to the entire universe.

Einstein envisioned the universe as being a sphere of some sort, which was expanding. We live on the surface of this sphere. This is called the Big Bang theory. But if we apply the quantum theory to the entire universe, this means that the universe, like the electron, must exist in many parallel states.

So if you try to apply superposition to the entire universe, then *you necessarily wind up with parallel universes,* just as Everett predicted. In other words, the starting point of quantum mechanics is that the electron can be in two states at the same time. When we apply quantum mechanics to the entire universe, it means that the universe must also exist in parallel states, i.e., in parallel universes. So parallel universes are unavoidable.

Thus, parallel universes necessarily emerge when you try to describe the entire universe in quantum terms. Instead of parallel electrons, now we have parallel universes.

But this leaves open the next question: Can we visit these parallel universes? Why don't we see this infinite collection of parallel universes, some of which might resemble our own, while others might be bizarre and preposterous? (And one question I often get is: Does this mean that Elvis is still alive in another universe? Modern science says: perhaps.)

Parallel Universes in Your Living Room

Nobel laureate Steve Weinberg once explained to me how to mentally get your head around the many worlds theory so your mind doesn't explode. Imagine, he said, sitting quietly in your living room, with radio waves from various radio stations around the world filling the air. In principle, there are hundreds of signals from various radio stations in your living room. But your radio is only tuned to one frequency; it can only pick up one station, because you are no longer vibrating in sync with other radio stations. In other words, your radio has "decohered" from the other radio waves filling up your living room. Your living room is full of different radio stations, but you cannot hear them because you are not tuned into them, or coherent with them.

Now, he told me, replace the radio waves with quantum waves of electrons and atoms. In your very living room, there are the waves of parallel universes, i.e., the waves of dinosaurs, aliens, pirates, volcanoes. However, you cannot interact with them anymore because you have decohered from them. You no longer vibrate in unison with the waves of dinosaurs. These parallel universes are not necessarily in outer space or another dimension. They could be in your living room. So it is possible to enter a parallel universe, but, when you calculate the chances of this happening, you find that you have to wait an astronomical amount of time for this to occur.

People who have passed away in our universe may actually be alive and well in a parallel one, right in our living room. But it is nearly impossible to interact with them because we are no longer coherent with them. So Elvis may be alive, but he is belting out his hits in another parallel universe.

The probability of entering into these parallel universes is nearly zero. The key word is "nearly." In quantum mechanics, everything is reduced to a probability. For example, for our PhD students, we sometimes ask them to calculate the probability that they will wake up on Mars the next day. Using classical physics, the answer is never, because we cannot escape the gravitational barrier keeping us rooted to the earth. But in the quantum world, you can calculate the probability that you "tunnel" your way past the gravitational barrier and wake up on Mars. (When you actually do the calculation, you find you have to wait longer than the lifetime of the universe for this to happen, so, most likely, you will wind up in your bed tomorrow.)

David Deutsch takes these mind-boggling concepts seriously. Why are quantum computers so powerful? he asks. Because the electrons are simultaneously calculating in parallel universes. They are interacting and interfering with each other via entanglement. So they can quickly outrace a traditional computer that computes in only one universe.

To demonstrate this, he takes out a portable laser experiment that he keeps in his office. It simply consists of a sheet of paper with two holes in it. He shines a laser beam through both holes and finds a beautiful interference pattern on the other side. This is because the wave has passed through both holes simultaneously, and has interfered with itself on the other side, giving rise to an interference pattern.

This is nothing new.

But now, he says, gradually reduce the intensity of the laser beam to almost zero. Eventually you have not a wave front, but just a single photon passing through both holes. But how can a single photon of light pass through both holes simultaneously?

In the usual Copenhagen interpretation, before you measure the photon, it actually exists as the sum of two waves, one for each hole. Isolating a single photon has no meaning until you measure it. Once you measure it, then you know which hole it went through.

Everett did not like this picture, because it meant that you could never answer the question: Which hole did the photon enter before we measured it? Now apply this to electrons. In Everett's many worlds theory, the electron is a point particle that indeed went through just one hole, but there was another twin electron in a parallel universe that went through the other hole. These two electrons, in two different universes, then interacted with each other via entanglement to alter the trajectory of the electron to create the interference pattern.

In conclusion, a single photon can pass through only one slit, but it can still create an interference pattern because the photon can interact with its counterpart moving in a parallel universe.

(Remarkably, physicists even today argue over various interpretations of the "collapse" of the wave function. But today, not just physicists, but schoolchildren are enamored of this idea, because many of their favorite comic book superheroes live in the multiverse. When their favorite superhero is in a jam, sometimes their counterpart in a parallel universe comes to the rescue. So quantum physics has become a hot topic even for kids.)

Summary of Quantum Theory

Let's now summarize all the bizarre features of the quantum theory that make quantum computers possible.

1. *Superposition.* Before you observe an object, it exists in many possible states. So an electron can be in two places at the same time. This vastly increases the power of a computer, since you have more states to calculate with.

2. *Entanglement.* When two particles are coherent and you separate them, they can still influence each other. This interaction takes place instantly. This allows atoms to communicate with each other, even when separated. This means that computer power grows exponentially as more and more qubits are added that can interact with each other, far faster than ordinary computers.

3. *Sum over paths.* When a particle moves between two points, it sums over all possible paths connecting these two points. The most likely path is the classical, nonquantum path, but all these other paths also contribute to the final quantum path of the particle. This means that even paths which are extremely unlikely may become real. Perhaps the paths of molecules that created life became real because of this effect, making life possible.

4. *Tunneling.* When faced with a large energy barrier, normally a particle fails to penetrate it. But in quantum mechanics, there is a small but finite probability that you can "tunnel" or penetrate through the barrier. This might be why the complex chemical reactions of life can proceed at room temperature, even without vast amounts of energy.

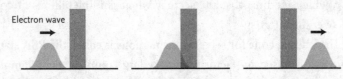

Figure 7: Tunneling
Normally, a person cannot walk through a brick wall. But in quantum mechanics, there is a small but finite probability that you can "tunnel" right through it. In the subatomic world, tunneling is commonplace and may explain how exotic chemical reactions that make life possible can take place.

Shor's Breakthrough

Up until the 1990s, quantum computers were still largely a plaything for theoreticians. They existed in the minds of a small but brilliant core of scientists, true believers, and academics.

But the work of Peter Shor at AT&T in the early 1990s changed everything. Far from being a minor footnote talked about casually at water coolers, quantum computers suddenly were on the agenda of major governments around the world. Security analysts, who may have little need for a physics background, were now being asked to decipher the mysteries of the quantum theory.

Everyone who watches a James Bond movie knows that the world, with so many competing and even hostile national interests, is full of spies and secret codes. This may be a Hollywood exaggeration, but the crown jewels of these security agencies are the codes they use to protect their most valuable national secrets. We recall that Turing's success breaking the Nazi Enigma code was a historic turning point, helping shorten the length of the war and altering the course of human history.

Up to then, work on quantum computers was highly speculative and was the domain of the most esoteric electrical engineers. But Shor showed that it is possible for a quantum computer to break any digital code currently in use, thereby jeopardizing the world economy, which requires absolute secrecy when sending billions of dollars over the internet.

The leading code for secret transmissions is called the RSA standard and is based on factoring a very large number. For example, start with two numbers, each 100 digits long. If you multiply them together, you can get a number approaching 200 digits. Multiplying two numbers is an easy task.

But if someone gave you this 200-digit number to start with, and asked you to factorize it (find the two numbers that multiply together to make it), it might take centuries or more to do this with a digital computer. This is called a trapdoor function. In one direction, when multiplying two numbers, the trapdoor function is trivial. But in the other direction, it's very difficult. Both classical and quantum computers can factorize a large number. In fact, a classical computer can, in principle, compute anything that a quantum computer can compute, and vice versa, but if the data is too complex it can overwhelm classical computers.

The key advantage of a quantum computer is time. Although both classical and quantum computers can perform certain tasks, the time it takes classical computers to crack a difficult problem may make it totally impractical.

So the time it takes for a classical computer to factorize a large number is prohibitively large, making it impractical to break into our secrets. But a quantum computer can crack the code after a given amount of time, which is still large but maybe just small enough to be practical.

So when hackers try to break into your computer, the computer will ask them to factorize a number, perhaps with 200 digits. Given how long this process will take, the hackers may simply give up. But if you want the intended recipient to read the transmission, all you need to do is give them the two smaller numbers beforehand. Then they can easily unlock the computer program protecting the message.

The RSA algorithm seems secure for now, but in the future it may be possible using quantum computers to factorize this 200-digit number.

To see how this works, let us examine Shor's algorithm. Over the centuries, mathematicians have devised algorithms to help them factorize a number into its prime factors, i.e., numbers whose own factors are only one and themselves. For example, $16 = 2 \times 2 \times 2 \times 2$, since 2 can only be divided by itself and 1.

Shor's algorithm starts with these standard techniques known to classical mathematicians to factorize an arbitrary number. Then, toward the end of the process, one performs what is called a Fourier transform. This involves summing over a complex factor, so the calculation proceeds normally. But in the quantum case, we have to sum over many, many more states, so instead we have to perform a quantum Fourier transform. The end result shows that the calculation can be done in record time because we have many more states to work with.

In other words, both a classical computer and quantum computer factorize in much the same way, except the quantum computer computes over many states simultaneously, which greatly speeds up the process.

Let N represent the number we wish to factorize. For an ordinary digital computer, the amount of time it takes to factorize a number grows exponentially, like $t \sim e^N$, times some unimportant factors. So the calculation time can rapidly rise to astronomical heights, comparable to the age of the universe. This makes the factorization of a large number possible but highly impractical on a conventional computer.

But if we do the same calculation using a quantum computer, the time to factorize only grows like $t \sim N^n$, i.e., like a polynomial, because quantum computers are astronomically faster than a digital computer.

Defeating Shor's Algorithm

Once the intelligence community became fully aware of the implications of this breakthrough, it began to take steps to confront it.

First, the National Institute of Standards and Technology, which sets technical standards for the U.S. government, issued a statement about quantum computers, saying that the real threat from quantum computers is still years away. But the time to start thinking about them is here and now. In the future, it might be too late to retool an entire industry on a moment's notice once quantum computers start cracking your codes.

Next, it suggested a simple measure that can be taken by companies to partially confront this threat. The easiest way to deal with Shor's algorithm is to simply increase the number that has to be factorized. Eventually quantum computers may still be able to break into a modified RSA code, but this will delay any hacker and perhaps make it prohibitively expensive to do so.

But the most direct way to address this problem is to devise more sophisticated trapdoor functions. The RSA algorithm is too simple to stump a quantum computer, so the NIST memo mentioned several new algorithms that were more complex than the original RSA code. However, these new trapdoor functions are not easy to implement. It remains to be seen if they can stop quantum computers.

The government urged companies and agencies to take measures to prepare for this digital cataclysm. In the U.S., guidelines were issued by the NIST on how to lay the groundwork to fight off this new threat to national security.

But if worse comes to worst, governments and large institutions may use the last resort, which is using quantum cryptography to defeat quantum computers, i.e., use the power of the quantum against itself.

Laser Internet

In the future, top secret messages may be sent on a separate internet channel carried by laser beams, not electrical cables. Laser beams are polarized, meaning that the waves vibrate in only one plane. When a criminal tries to tap into the laser beam, this changes the direction of polarization of the laser, which is immediately detected by a monitor. In this way, you know, by the laws of the quantum theory, that someone has tapped into your communication.

So if a criminal tries to intercept a transmission, it will inevitably set alarm bells ringing. It does require, however, a separate internet based on lasers to carry the most important national secrets, which would be an expensive solution.

This might mean that, in the future, there could be two layers to the internet. Some organizations, like banks, large corporations, and governments, may pay a premium to send messages on a laser-based internet, which is guaranteed to be secure, while everyone else will use the ordinary internet, which does not have this extra costly layer of protection.

This problem of security is also leading to a new technology called quantum key distribution (QKD), which transfers encryption keys using entangled qubits, so that one can detect immediately if someone is hacking into your network. Already, Japan's Toshiba company has predicted that QKD may generate up to $3 billion in revenue by the end of this decade.

So for now, it's a waiting game. Many are hoping the threat has been exaggerated. But that hasn't stopped the world's leading corporations from engaging in a race to see which technology will dominate the future.

Looking beyond the cyber threat, there are entirely new worlds to conquer with quantum computers, and companies are now jostling to gain the upper hand with this exciting, emerging technology.

The winner may be able to shape the future.

CHAPTER 5

THE RACE IS ON

Some of the biggest names in Silicon Valley are now placing their bets on which horse will win this race. It's too early to tell at this point who that might be, but what is at stake is nothing less than the future of the world economy.

To understand how the race is shaping up, it is important to realize that there is more than one computer architecture that will work. Recall that the Turing machine is based on general principles that can be applied to a wide range of technologies. Thus, it's possible to make a digital computer out of water pipes and valves. The essential ingredient is a system that can carry digital information characterized by a series of 0s and 1s, and a way to process this information.

Similarly, quantum computers can also have a wide range of possible designs. Basically, any quantum system that can superimpose states of 0s and 1s and entangle them so that they can process this information can become a quantum computer. Electrons and ions that spin up or down could serve this purpose, or polarized photons that spin clockwise or counterclockwise. Since the quantum theory governs all matter and energy in the universe, there are potentially thousands of ways to build a quantum computer. In a lazy afternoon, a physicist may dream up scores of ways of representing the superposition of 0s and 1s to create an entirely new quantum computer.

So what do those various designs look like, and what are the advantages and disadvantages of each? As we saw, companies and governments are investing billions in this technology, and their choice of design may influence who will come to dominate this race. So far, IBM is leading the pack with 433 qubits, but like a horse race, the exact rankings can change at any time.

Name	Producer	Qubits
Osprey	IBM	433
Jiuzhang	China	76
Bristlecone	Google	72
Sycamore	Google	53
Tangle Lake	Intel	49

As we go to press, IBM released the 433-qubit Osprey quantum computer and will deploy the 1,121-qubit Condor quantum computer in 2023. Dario Gil, IBM senior vice president and head of its research division, says, "We believe that we will be able to reach a demonstration of quantum advantage—something that can have practical value—within the next couple of years. That is our quest." In fact, IBM has publicly stated that its goal is to eventually build a million-qubit quantum computer.

So how does their industry-leading design work, and what does the competition look like?

1. Superconducting Quantum Computer
At present, the superconducting quantum computer has set the bar for computing power. Back in 2019, Google was first out of the

gate, announcing that it had achieved quantum supremacy with its Sycamore superconducting quantum computer.

However, IBM was not far behind, and later surged ahead with its Eagle quantum processor, which broke the 100-qubit barrier in 2021 and has since developed the 433-qubit Osprey processor.

Figure 8: Quantum computer
A quantum computer like the one shown here often resembles a large chandelier. Most of the complex hardware in this picture consists of the pipes and pumps necessary to cool the core to near absolute zero. The actual heart of a quantum computer may be as small as a quarter, located toward the bottom of the picture.

Superconducting quantum computers have a great advantage: they can use off-the-shelf technology pioneered by the digital computer industry. Silicon Valley companies have had decades to master the art of etching tiny circuits on silicon wafers. Within each chip, it's possible to represent the numbers 0 and 1 by the presence or absence of electrons in the circuit.

The superconducting quantum computer relies on this technology as well. By bringing the temperature down to a fraction of a degree above absolute zero, the circuits become quantum mechanical, i.e., they become coherent, so the superposition of electrons is undisturbed. Then, by bringing various circuits together, one can entangle them so that quantum calculations are possible.

But the disadvantage of this approach is that an elaborate array of tubes and pumps is necessary to cool the machine down. This also raises the cost and introduces the possibility of new complications and errors. The slightest vibration or impurity can break the coherence of the circuits. Someone sneezing nearby can ruin an experiment.

Scientists measure this sensitivity by something called coherence time, i.e., the length of time that atoms remain coherently vibrating together. In general, the lower the temperature, the slower the motion of the atoms in the environment, and the longer the coherence time. Cooling the machines to temperatures even lower than those in outer space maximizes the coherence time.

However, because it is impossible to actually reach absolute zero, errors will inevitably creep into the calculation. While an ordinary digital computer does not have to worry about this, it becomes a major headache for the quantum computer. It means that you cannot entirely trust the results. This could be a serious problem if billions of dollars in transactions are at stake.

One solution to this problem is to back up each qubit with a collection of qubits, which creates redundancy and reduces the errors of the system. For example, let's say a quantum computer does a calculation with three qubits backing up each qubit, and produces the string of numbers 101; since the values do not all match, then most likely the center digit is wrong and should be replaced by a 1. Redundancy can reduce the errors in the final result, but at the cost of vastly increasing the number of qubits in the system.

It has been suggested that perhaps 1,000 qubits might be necessary to back up just one qubit, so that this collection of qubits can correct for errors creeping into the calculation. But this means that, for a 1,000-qubit quantum computer, you need a million qubits.

This is a huge number that will push the technology to the limit, but Google estimates that a million-qubit processor may be attainable within ten years.

2. Ion Trap Quantum Computer

Yet another contender is the ion trap quantum computer. When you take an electrically neutral atom and strip off some electrons, you get a positively charged ion. An ion can be suspended in a trap consisting of a series of electric and magnetic fields, and when multiple ions are introduced they vibrate as coherent qubits. For example, if the electron axis spins up, then the state is a 0. If it spins down, it is a 1. So the result due to the strange effects of the quantum world is a superimposed mixture of two states.

Then microwave or laser beams can hit these ions, flipping them and causing them to change state. So these beams act like a processor,

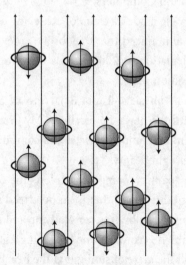

Figure 9: Ion quantum computer
Atoms can spin like a top and become aligned by a magnetic field. If an atom spins upward, it can represent the number 0. If it spins downward, it can be a 1. But atoms can also exist in a superposition of these two states. A calculation is made by hitting these atoms with a laser, which can flip the spins and interchange the 0s and 1s, thereby making a calculation.

turning one configuration of atoms into another, just as a CPU in a digital computer flips transistors between on and off states.

So this is perhaps the most transparent way to see how a quantum computer emerges from a collection of random electrons. Honeywell is one of the leading proponents of this model.

In an ion trap quantum computer, the atoms are kept in a near-vacuum state, suspended by a complex array of electric and magnetic fields that can absorb random motions. The coherence time can therefore be much longer than in a superconducting quantum computer, and the ion computer can actually operate at higher temperatures than its rivals. But one problem, however, is scaling, i.e., when you try to increase the number of qubits. Scaling is quite difficult, since you have to continually readjust the electric and magnetic fields to maintain coherence, which is a complex process.

3. Photonic Quantum Computers

Soon after Google made its claim of achieving quantum supremacy, the Chinese announced that they broke an even larger barrier, performing a calculation in 200 seconds that would take a digital computer half a billion years.

When quantum physicist Fabio Sciarrino of Sapienza University in Rome heard the news, he recalled, "My first impression was, Wow!" Their quantum computer, instead of computing on electrons, computes on laser light beams.

The photonic quantum computer exploits the fact that light can vibrate in different directions, that is, in polarized states. For example, a light beam may be vibrating vertically up and down, or perhaps sideways, left and right. (Anyone who buys sunglasses with polarized lenses to diminish the glare of sunlight at the beach takes advantage of this. For example, your polarized glass may have a series of parallel grooves in the vertical direction, which blocks sunlight vibrating in the horizontal direction.) So the number 0 or 1 can be represented by light vibrating in different polarized directions.

The photonic quantum computer starts by firing a laser beam at a beam splitter, which is just a finely polished piece of glass, at a forty-five-degree angle. Hitting it, the laser beam fissions in two, with half going forward and the other half being reflected sideways. The important point here is that the two beams are each coherent, vibrating in unison with each other.

Then the two coherent beams can hit two polished mirrors, which then reflect the two beams back to a common point, where the two photons get entangled with each other. In this way, we can create a qubit. Thus, the resulting beam is now a superposition of two entangled photons. Now imagine a tabletop consisting of perhaps hundreds of beam splitters and mirrors, which entangle a series of coherent photons together. That is how the optical quantum computer performs its miraculous feats. The Chinese photonic computer was able to calculate with 76 entangled photons moving in 100 channels.

But photonic computers have one serious drawback: they are an ungainly collection of mirrors and beam splitters that can easily fill up a large space. For each problem, you have to rearrange the complex collection of mirrors and beam splitter into a different position. It is not an all-purpose machine that you can program to perform instant calculations. After each calculation, you have to tear it down and rearrange the components precisely, which is time-consuming. Furthermore, because photons don't interact easily with other photons, it is difficult to create qubits of increasing complexity.

However, there are several benefits to using photons rather than electrons for a quantum computer. While electrons react strongly with ordinary matter because they are charged (and hence disturbances from the environment can be quite large), photons are not charged, and hence encounter much less noise from the environment. In fact, light beams can pass right through other light beams with minimal disruptions. Photons are also much faster than electrons, traveling ten times the speed of electrical signals.

But the great advantage of the photonic computer, which may

eventually outweigh other factors, is that it can operate at room temperature. There are no expensive pumps and tubing necessary to drop the temperature down to near absolute zero, which can quickly drive up the cost.

Because photonic computers operate at room temperature, their coherence time is quite short. But this is compensated for by the fact that the laser beams can have high energy, so the calculations can be done much faster than the coherence time, so the molecules in the environment appear as if they were moving in slow motion. This reduces the amount of errors created by interactions with the environment. In the long haul, the advantages of lower error rates and reduced costs may outpace those of other designs.

More recently, a Canadian start-up called Xanadu has introduced its photonic quantum computer, which has a distinct edge. It is based on a tiny chip (not a tabletopful of optical hardware) that manipulates infrared laser light through a microscopic maze of beam splitters. Unlike the Chinese design, the Xanadu chip is programmable and its computer is available on the internet. However, it only has eight qubits, and still requires some superconducting freezers. But, as Zachary Vernon of Xanadu says, "For a long time, photonics was considered an underdog in the quantum computing race. . . . With these results . . . it's becoming clear that photonics is not an underdog, but in fact one of the leading contenders." Time will tell.

4. Silicon Photonic Computers

In 2016, a new company joined the race and caused considerable controversy. PsiQuantum, a brand-new start-up, convinced investors of its silicon photonic computer design and shocked Wall Street by amassing a staggering $3.1 billion valuation. It did this without ever producing a prototype or demonstration project showing that it actually works.

The big advantage of silicon photonic computers would be that they can use the tried-and-true methods perfected by the semiconductor industry. In fact, PsiQuantum is in a joint partnership with GlobalFoundries, which is one of the top three most advanced chip makers in the world. This joint venture with an established high-tech company gave this young enterprise immediate recognition on Wall Street.

One reason why PsiQuantum has generated so much media attention is because it laid out the most ambitious plan for the future yet. They claim that, by the middle of this century, they will create a million-qubit silicon optical computer that will have practical applications. They feel that their competitors, which have focused on quantum computers with about 100 qubits, are much too conservative because they concentrate on small, incremental advances. They hope to take giant leaps into the future, bypassing their more cautious and timid rivals.

One of the keys to their program is the dual nature of silicon. Not only can silicon be used to make transistors and hence control the flow of electrons, it can also be used to transmit light, since it is transparent to certain frequencies of infrared radiation. This dual nature is crucial to entangling photons.

One big selling point is that they can address the problem of error correction. Since errors creep into any calculation because of interactions with the environment, you want redundancy built into the system by creating redundant qubits. With a million qubits, they feel that they can begin to control these errors, so that real practical calculations can be done on the computer.

5. Topological Quantum Computers

The dark horse in this race is the Microsoft design, which uses topological processors.

As we've seen, one major problem facing several of the previous designs is that the temperature must be kept near absolute zero. But according to quantum theory, there is another way besides ion traps and photonic systems to create a quantum computer. A system can remain stable at room temperature if it maintains some special topological properties that are always preserved. Think of a circular piece of rope with a knot in it. If you are not allowed to cut the rope, then no matter how hard you try, the knot cannot be removed. The topology of the rope (the shape, in this case the knot) cannot be changed by any manipulation besides cutting it. Similarly, physicists have tried to find physical systems that preserve the topology of the system no matter what the temperature is. If found, it would greatly reduce the cost and increase the stability of a quantum computer. With such a system, coherent qubits could be created out of these topological configurations.

In 2018, physicists at the Delft University of Technology in the Netherlands announced they had discovered a material with these properties, indium antimonide nanowires. This material arose out of a complex series of interactions of many constituent substances, and was therefore "emergent." It was called a Majorana zero mode quasiparticle. The media touted it as a magic material that would be stable even at room temperatures. Microsoft even generously opened its checkbook and began to set up a new quantum laboratory on campus.

Just as it looked like a breakthrough of the highest magnitude had occurred, another group announced that they could not duplicate the result. Upon close examination, the Delft group announced that perhaps they had rushed to interpret their results, and they retracted their paper.

The stakes are so high that even physicists begin to believe their press releases. However, other topological objects are still being studied, such as anyons, so this approach is still considered viable.

6. D-Wave Quantum Computers

There is currently one last type of quantum computing called quantum annealing, being pursued by the D-Wave company, based in Canada. Though it does not use the full power of quantum computers, D-Wave claims it can produce machines that can reach 5,600 qubits, far beyond the number found in other competing designs, and has plans to offer computers with more than 7,000 qubits in a few years. So far a number of high-profile companies have purchased D-Wave computers, which are being sold on the open market for between $10 and $15 million. These companies include Lockheed Martin, Volkswagen, Japan's NEC, Los Alamos National Laboratory, and NASA. Apparently, D-Wave quantum computers excel in one area, optimization. Companies that are interested in optimizing certain parameters of their business (such as reducing waste, maximizing efficiency, increasing profit) have invested in this technology. D-Wave computers can optimize data by using magnetic and electric fields to manipulate currents flowing in superconducting wires, until they reach the lowest energy state.

In summary, there is intense competition among corporations and even governments to get a head start on this new technology. The rate of progress in this field has been astounding. Every major computer company has their own quantum computer program. Prototypes are already proving their worth and are even being sold on the marketplace.

But the next big challenge is for quantum computers to solve real-world practical problems that can alter the trajectory of entire industries. Scientists and engineers are focusing on problems that are far beyond the reach of digital computers. The goal is to apply quantum computers to solve the biggest problems in science and technology.

One focus of research is to uncover the quantum mechanics behind the origin of life, which will help unravel the mystery of photosynthesis, feed the planet, provide society with energy, and cure incurable diseases.

PART II

QUANTUM COMPUTERS AND SOCIETY

CHAPTER 6

THE ORIGIN OF LIFE

E very culture has its cherished mythology about the beginning of life. People have often wondered what could possibly explain the glorious richness and diversity on earth. In the Bible, for example, God created the heavens and earth in six days. He created man in His image out of dust, and then breathed life into him. He created all the plants and animals to be ruled by us.

In Greek mythology, there was only formless Chaos and the void in the beginning. But out of this enormous emptiness was born the gods, such as Gaea, goddess of the earth, Eros, the goddess of love, and Ether, the god of light. The union of Gaea and Uranus, the god of the night sky, then produced the creatures that populated the earth.

The origin of life is perhaps one of the greatest mysteries of all time. This question has dominated religious, philosophical, and scientific discussions like no other. Throughout history, many of the deepest thinkers believed that there was a mysterious "life force" that could animate the inanimate. Many scientists, in fact, believed in something called spontaneous generation, that life could magically arise by itself out of inanimate matter.

In the 1800s, scientists were able to piece together many of the clues about where life comes from. Careful experiments by Louis Pasteur and others showed decisively that life could not be spontaneously generated, as was commonly believed. He showed that by boiling water, one could create a sterile environment that would prevent organisms from developing spontaneously.

Even today, there are many gaps in our understanding of how life first originated on the earth almost 4 billion years ago. In fact, digital computers are useless when analyzing the fundamental biological and chemical processes at the atomic level that might shed light on this problem. Even the simplest molecular process can quickly overwhelm the capacity of a digital computer. However, quantum mechanics may help explain many of these gaps and unravel the mysteries of life. Quantum computers are ideally suited for this problem and are now beginning to uncover some of the deepest secrets of life at the molecular level.

Two Breakthroughs

Two monumental breakthroughs occurred in the 1950s that have set the agenda for further research in the origins of life. The first occurred in 1952, when a graduate student, Stanley Miller, working under Harold Urey at the University of Chicago, did a simple experiment. He began with a flask of water and then added a toxic brew of chemicals including methane, ammonia, water, hydrogen, and other substances, which he thought mimicked the harsh atmosphere of the early earth. In order to add energy to the system (perhaps mimicking lightning bolts or UV radiation from the sun), he added a small electrical spark. And then walked away from the experiment for a week.

When he came back, he found a red liquid inside the flask. Upon careful examination, he realized that the coloration was caused by amino acids, which are the basic constituents of the proteins of our body. In other words, the basic ingredients of life formed without any outside interference.

Since then, this simple experiment has been repeated and modified hundreds of times, giving scientists a revealing look into the ancient chemical reactions that may have spawned life. One can imagine, for example, that toxic chemicals found in hydrothermal vents at the bottom of the oceans might have provided the basic elements needed to create the first chemicals of life and that these volcanic vents might have then supplied the energy to turn those chemicals into the amino acids necessary for life. Indeed, some of the most primitive cells on earth are found near these underwater volcano vents.

Today, we realize how easy it is to create the building blocks of life. Amino acids have been found in distant gas clouds many light-years away, or in the interior of meteorites from outer space. Carbon-based amino acids may form the seeds of life throughout the universe. And all of this because of the simple bonding properties of hydrogen, carbon, and oxygen, as predicted by the Schrödinger equation.

Thus, it should be possible to apply quantum mechanics to find, step by step, the quantum processes that originated life on earth. Elementary quantum theory helps us understand why the Miller experiment was so successful, and it may point the way toward deeper discoveries in the future.

First, using quantum mechanics, one can calculate the energy necessary to break the chemical bonds of methane, ammonia, etc., to create amino acids. Equations from quantum mechanics show us that an electrical spark like the one in the Miller experiment has enough energy to do this. Further, it shows us that if the activation energy necessary to break these chemical bonds was somehow much larger, then life would never have emerged.

Second, we see that carbon has six electrons. Two sit in the first-level orbital, and the remaining four sit individually in the four spaces of the second-level orbitals. This leaves room for four chemical bonds. An element with four bonds is rare among the chemicals in the periodic table. But the rules of quantum mechanics allow this structure to create long, complex chains of carbon, oxygen, hydrogen, and nitrogen, thereby creating the amino acids.

Third, these chemical reactions take place in water, H_2O, which acts like a melting pot where different molecules meet and form more complex chemicals. Using quantum mechanics, one finds that the water molecule is shaped like the letter L, and one can calculate that the two hydrogen atoms make a 104.5-degree angle with respect to each other. This, in turn, means that the water molecule has a net electrical charge distributed unevenly around the molecule. This electric charge is large enough to break apart the weak bonds of other chemicals, so water can dissolve many chemicals.

Thus, we see that basic quantum mechanics can create the conditions for life. But the next question is, can we go beyond the Miller experiment and see if the quantum theory can create DNA? And beyond that, can quantum computers be applied to the human genome to decipher the secrets of disease and aging?

What Is Life?

The second breakthrough came directly from quantum mechanics. In 1944, Erwin Schrödinger, already famous for his wave equation, wrote a seminal book, *What Is Life?* In it, he made the astonishing claim that life itself is a by-product of quantum mechanics, and that the blueprint of life is encoded in an unknown molecule. In an era when many scientists still believed that a mysterious "life force" animated all living matter, he made the assertion that life can be explained by an application of quantum physics. By examining solutions of his wave equation, he conjectured, life could arise from pure mathematics, in the form of a code handed down through this mystery molecule.

It was an outrageous idea. But two young scientists, physicist Francis Crick and biologist James Watson, saw this as a challenge. If the basis of life could be found in a molecule, then their task would be to find this molecule and prove that it carried the code of life.

"From the moment I read Schrödinger's *What Is Life?*, I became polarized towards finding out the secret of the gene," recalls Watson.

They reasoned that the molecule of life, as envisioned by Schrödinger, must be hidden in the genetic material of the nucleus of the cell, much of which is composed of a chemical called DNA. But since organic molecules like DNA are so tiny (even smaller than the wavelength of visible light), they are invisible, their task seemed daunting. They chose an indirect method, using the quantum-theory-based process of X-ray crystallography to find this mythical molecule.

X-rays, unlike visible light, can have a wavelength as small as atoms. If X-rays are then shot through a crystal consisting of trillions upon trillions of molecules arranged in some lattice, the scattered X-rays form a distinct interference pattern, which can be photographed. Upon close examination, a trained physicist can study the photographic plates to determine what crystallized pattern created these images.

Glancing at the X-ray photographs of DNA taken by Rosalind Franklin, Crick and Watson saw a pattern that they recognized must be created by a double helix. Knowing that the overall structure of DNA was a double helix, like two staircases that wrap around each other, they were able to piece together the entire structure of DNA, atom for atom.

Quantum mechanics gave them the angles formed by bonds containing carbon, hydrogen, and oxygen atoms. So, like children building a Lego set, they were able to reconstruct the complete atomic structure of DNA and explain how it was able to make copies of itself and provide the instructions for all biological development.

This, in turn, has altered the very nature of biology and medicine. In the previous century, Charles Darwin was able to sketch the Tree of Life, with all the branches representing the rich diversity of forms. This huge Tree of Life was set into motion by just one molecule. And, as envisioned by Schrödinger, all of this can be deduced as a consequence of mathematics.

When they unraveled the DNA molecule, they found that it was built out of four clusters of atoms, called nucleic acids. These four nucleic acids, called A, C, T, and G, are arranged in a linear sequence to form two long parallel lines, which are then intertwined like a staircase to create the DNA molecule. (A strand of DNA is invisible, but if it were unfurled, this single molecule would be about six feet long.) When it is time to reproduce, the two strands of DNA unwind and separate into two strands of nucleic acids. Then each strand acts like a template, grabbing other atoms in the right order so that each single strand becomes a double strand once again. In this way, life can reproduce itself.

We now had the architecture by which to create the DNA molecule using the mathematics of the quantum theory. But determining the basic shape of the DNA molecule was, in some sense, the easy part. The hard part is to decipher the billions of codes hidden within the molecule.

It's as if you are trying to understand music, and you finally learned how to plunk a few notes on a piano keyboard. But that does not make you a Mozart. Learning a few notes is just the beginning of a long journey.

Physics and Biotechnology

One person who spearheaded this effort to sequence all our genes was Harvard biochemist and Nobel laureate Walter Gilbert. When I interviewed him, he admitted to me that this field was not in his original game plan. In fact, he started working at Harvard as a professor of physics, studying the behavior of subatomic particles created in powerful accelerators. Working on biology was the furthest thing from his mind.

But he began to change his thinking. First, he realized how difficult it would be to get tenure at Harvard with so much competition. The field of particle physics had many bright researchers with whom he had to compete. As it turned out, his wife was working for James Watson, whom he had met earlier while at Cambridge University, so he became familiar with the pioneering work being done in the new field of biotechnology, which was exploding with ideas and discoveries. Intrigued, he found himself splitting his time between the arcane equations of elementary particles and getting his hands dirty with biology.

So he made the biggest gamble of his career.

As a professor of physics, he made a huge jump, switching from theoretical elementary particle physics to biology. But the gamble paid off, because in 1980 he won the Nobel Prize in Chemistry. Among other achievements, he was one of the first to develop a rapid technique to read the DNA molecule, gene for gene.

Coming from a physics background actually helped him. Traditionally, most biology departments were filled with people who specialized in one animal or plant. Some would spend their lives finding and giving names to newly discovered species. But all of a sudden, breakthroughs were being made by quantum physicists using advanced calculus. Being fluent in the abstruse language of quantum mechanics helped him make the breakthrough that altered our understanding of the molecular basis of life.

He then helped build momentum for the Human Genome Project. In 1986, when speaking at Cold Spring Harbor in New York, he gave an estimate for the cost of this ambitious, unprecedented endeavor: $3 billion. "The audience was stunned," recalled Robert Cook-Deegan, author of *The Gene Wars*. "Gilbert's projections provoked an uproar." This, many people felt, was an impossibly low number. When he made that startling prediction, only a handful of genes had been sequenced. Many scientists even thought that the human genome would forever be beyond reach.

But that number became the budget that Congress approved for the Human Genome Project. The technology was advancing so rapidly that the project was completed ahead of schedule and under budget, which is unheard of in Washington. (I asked him how he arrived at that number. He realized there were 3 billion base pairs in our DNA, and he estimated that it would eventually cost $1 to sequence one base pair.)

Gilbert even made the prediction that, in the future, "You'll be able to go to a drugstore and get your own DNA sequence on a CD, which you can then analyze at home on your Macintosh . . . [you] will be able to pull a CD out of [your] pocket and say, 'Here's a human being; it's me!'"

One person who was deeply influenced by all this is Francis Collins, the former director of the National Institutes of Health. He is one of the most influential doctors in medicine today. Millions of people have seen him on TV talking about the latest developments with the Covid-19 pandemic.

I asked Collins how he became interested in biology, despite starting out as a chemistry major. He confessed to me that biology always seemed so "messy," with so many arbitrary names for so many animals and plants. There was no rhyme or reason, he thought. In chemistry, he saw order, discipline, and patterns that could be studied and duplicated. So he taught physical chemistry, using the Schrödinger equation to explain the inner workings of molecules.

However, he eventually realized that he was in the wrong field. Physical chemistry was well established, with well-known principles and concepts.

Then he began to reconsider biology. While in biology scientists gave strange Greek names to obscure bugs and animals, the field of biotechnology was exploding with new ideas and fresh concepts. It was uncharted, virgin territory for newcomers.

He consulted with others, including Walter Gilbert, who told him how he had made the switch from elementary particle physics to DNA sequencing. He encouraged Collins to do the same.

So Collins took the plunge and never regretted it. He recalled, "I realized, 'Oh my gosh, this is where the real golden era is happening.' I was worried that I would be teaching thermodynamics to a bunch of students who absolutely hated the subject. Whereas what was going on in biology seemed like quantum mechanics in the 1920s. . . . I was completely blown away."

Very quickly, Collins made a name for himself. In 1989, he uncovered the gene mutation responsible for cystic fibrosis. He found that it is caused by the deletion of just three base pairs in your DNA (from ATCTTT to ATT).

Eventually, he became the top medical administrator in the country. But he brought his own personal style to Washington. He rode to work on his motorcycle. And he has never shied away from his personal religious beliefs. He even wrote a best-seller: *The Language of God: A Scientist Presents Evidence for Belief.*

Three Stages in Biotechnology

Gilbert and Collins, in some sense, represent some of the stages in the development of this field.

Stage One: Mapping the Genome

In Stage One, Walter Gilbert and others were able to complete the Human Genome Project, one of the most important scientific ventures of all time. However, the catalogue of the human genome is like a dictionary with 20,000 entries and no definitions. By itself, it is a monumental accomplishment, but also a useless one.

Stage Two: Determining the Function of the Genes

In Stage Two, Francis Collins and others have tried to fill in the definitions for these genes. By sequencing diseases, tissues, organs, etc., one is able to tediously compile the way in which these genes operate. It is a painfully slow process, but gradually the dictionary is being filled up.

Stage Three: Modifying and Improving the Genome

But now we are gradually entering Stage Three, when we can use this dictionary to become writers ourselves. This means using quantum computers to decipher how these genes operate at the molecular level, so that we can devise new therapies and create new tools to attack incurable diseases. Once we understand how they inflict their damage at the molecular level, we may be able to use that knowledge to devise new techniques to neutralize or cure these diseases.

Paradox of Life

In trying to trace the origin of life, we are still faced with a glaring paradox staring at us. How could random chemical events create the exquisitely complex molecules of life and in such a short period of time?

Geologists believe that the earth is 4.6 billion years old. For almost a billion years, the earth was molten and too hot to sustain life. Because of repeated meteor impacts and volcanic eruptions, the ancient oceans probably boiled off on several occasions, making life impossible. But by 3.8 billion years ago, the earth had gradually cooled down enough to allow oceans to form. Since DNA is believed to have originated around 3.7 billion years ago, this means that in a couple hundred million years, DNA suddenly got off the ground, complete with the chemical processes that allow it to use energy and reproduce.

Some scientists have said that they believe this is impossible. Fred Hoyle, one of the great pioneers in cosmology, believed that given how quickly DNA seems to have emerged, there was simply not enough time for life to have formed on earth, so it must have come from outer space. Rocks and gas clouds in deep space are known to contain amino acids, so perhaps life originated elsewhere.

This is called the panspermia theory and recently new evidence has reignited interest in it. By examining the mineral content and the tiny air bubbles trapped inside meteorites, one finds an exact match with the rocks found on Mars with our space probes. Of the 60,000 meteorites that have been discovered so far, at least 125 of them have been conclusively identified as having come from Mars.

For example, one meteor called ALH 84001 fell on the South Pole 13,000 years ago. It was probably blasted into space by a meteor impact 16 million years ago, and then drifted until finally landing on the earth. Microscopic analysis of the interior of the meteor shows evidence of some wormlike structures. (Even today, there is debate about whether these structures are ancient fossilized multicelled creatures, or a naturally occurring phenomenon.) If rocks can travel from Mars to earth, then why not DNA?

It is now believed that there may be scores of meteors drifting between Mars, Venus, the moon, and earth, where meteor impacts were large enough to send rocks into space and eventually land on another planet. One cannot rule out that DNA might have come from somewhere other than earth.

However, there is another explanation for this conundrum.

As we have seen, the quantum theory allows for several mechanisms to vastly accelerate a chemical process. The path integral method discussed earlier sums over all possible pathways in a chemical reaction, including even unlikely ones. Paths that are actually forbidden by the usual Newtonian rules may actually be possible with quantum mechanics. Some of these could lead to the creation of complex molecular structures.

We also know that enzymes can speed up chemical processes. They can bring chemicals together so they react quickly, and then lower the energy threshold so they can tunnel through the energy barrier. This means that even highly unlikely chemical reactions can become reality. Reactions that seemingly violate the conservation of energy may be allowed under the quantum theory.

So in other words, quantum mechanics could be the reason why life started so early on planet earth. With the coming of quantum computers, it is hoped that many of the missing gaps in our understanding of life may be solved.

Computational Chemistry and Quantum Biology

Whirlwind advances in quantum computers are giving birth to new sciences called computational chemistry and quantum biology. Finally, quantum computers are making it possible to create realistic models of molecules, allowing scientists the ability to see, atom for atom, nanosecond by nanosecond, how chemical reactions take place.

For example, think of using a cookbook to create a meal. It is convenient to simply follow instructions, step by step, but you have no idea of how the flavors and ingredients interact to create a delicious meal. If you deviate from the cookbook, then it's all trial and error and guesswork. It is time-consuming and leads to many dead ends. But this is pretty much how chemistry is done today.

Now imagine you could analyze all the ingredients at the molecular level. In principle it might be possible to create new, delicious recipes from first principles, knowing how the molecules all interact with each other. This is the hope of quantum computers, to be able to understand the interaction of genes, proteins, and chemicals at the molecular level.

Researcher Jeannette M. Garcia of IBM says, "As molecules get larger, they very quickly get out of the realm of what you can simulate with classical computers."

Elsewhere, Garcia has written that "predicting the behavior of even simple molecules with total accuracy is beyond the capabilities of the most powerful computers. This is where quantum computing offers the possibility of significant advances in the coming years." She points out that digital computers can only reliably calculate the behavior of just a couple of electrons. Beyond that, the calculation overwhelms any classical computer, unless drastic approximations are made.

She adds, "Quantum computers are now at the point where they can begin to model the energetics and properties of small molecules, such as lithium hydride—offering the possibility of models that will provide clearer pathways to discovery than we have now."

Linghua Zhu at Virginia Tech says, "The atoms are quantum, the computer is quantum, we're using quantum to simulate quantum. When we use classical methods, we always use approximations, but with a quantum computer, it's possible to exactly know how each atom is interacting with the others."

For example, think of an artist trying to paint a copy of the *Mona Lisa*. If you give the artist nothing but toothpicks, then the resulting picture is only a crude stick figure. Straight lines cannot capture the complexity of the human form. But if you give the artist a fine ink pen with different colors, then you can create a wealth of curved shapes that can create a reasonable copy of the famed painting. In other words, you need curved lines in order to simulate curved lines. Similarly, only a quantum computer can capture the complexity of quantum systems, such as the chemicals and the building blocks of life.

To see how this works, let us go back to the Schrödinger wave equation, mentioned in Chapter Three. Recall that we introduced a quantity called H (the Hamiltonian) that represents the total energy of the system being studied. This means that, for large molecules, that quantity consists of the sum of a large number of terms, such as:

- The kinetic energy of each electron and nuclei
- The electrostatic energy of each particle

- The interaction between all the various particles
- The effects of spin

If we are studying the simplest possible system—the hydrogen atom with just one electron and one proton—then this can be solved exactly in any first-year graduate course in physics. The derivation requires little more than third-year calculus. Still, for such a simple system, we get a veritable gold mine of results, such as the entire set of energy levels of the hydrogen atom.

But if we have just two electrons, representing the helium atom, things get complicated very fast, since we now have complex interactions between the two electrons. For three or more electrons, this rapidly spirals out of control for digital computers. Therefore, a large number of approximations have to be made to get reasonably accurate results. Quantum computers may be of benefit for this.

As an example, in 2020, it was announced that Google's Sycamore computer set a new record; it was able to accurately simulate a chain of twelve hydrogen atoms using twelve qubits.

"That's a result that we're pretty excited about, because this is more than double the number of qubits and the number of electrons as any prior quantum chemistry simulation, and it had the same level of accuracy," says Ryan Babbush, who was part of the team that set the new record.

The quantum computer was also able to model a chemical reaction involving hydrogen and nitrogen, even if one shifted the location of one of the hydrogen atoms. Babbush adds, "It shows that, in fact, this device is a completely programmable digital quantum computer that can be used for really any task you might attempt."

Garcia concludes, "Classically built computers simply cannot handle the level of complexity of substances as commonplace as caffeine." To her, the future is quantum.

But these initial accomplishments have only whetted the appetite of quantum scientists. They are eager to tackle even more ambitious projects, such as photosynthesis, which is the foundation for life on earth. The secret of how to take sunlight and create the bounty of fruits and vegetables we see all around us may one day be unraveled by quantum computers. So the next target may be photosynthesis, one of the most important quantum processes on the planet.

CHAPTER 7

GREENING THE WORLD

When I walk in a dense forest on a bright spring day, I can't help but be overwhelmed by the lush, vibrant green vegetation that surrounds me and the explosion of delicate blossoms everywhere I look. Wherever I gaze, I see this rainbow of vivid colors. I see life bursting out in all directions, with plants eagerly soaking up the sunlight and somehow converting that energy into all this abundance.

But I am also overwhelmed by the realization that I am witnessing a drama that has played out for over 3 billion years, a process that literally makes complex life on earth possible. What drives life on this planet is photosynthesis, the deceptively simple process by which plants convert carbon dioxide, sunlight, and water into sugar and oxygen. It's staggering to realize that photosynthesis creates 15,000 tons of biomass per second, which is responsible for covering the earth with green vegetation.

Life would be unimaginable without photosynthesis, yet remarkably, with all our advances in science, biologists are still not precisely sure how this vital process occurs. Some biologists believe that, because the capture of a photon of energy by photosynthesis is nearly 100 percent efficient, it must be quantum mechanical. (But if you calculate the overall efficiency for turning light into the final product of fuel and biomass, which requires a series of complex steps and intricate chemical reactions, then the final efficiency drops down to 1 percent.) If one day quantum computers can solve the secret of photosynthesis, then it would be possible to make photovoltaic cells with near-perfect efficiency, making the Solar Age a reality. We could also increase the yield from crops to feed a hungry planet. Perhaps photosynthesis could be modified so that plants could flourish even in harsh environments. Or, if one day we begin the colonization of Mars, it might be possible to modify photosynthesis so that vegetation can thrive on the Red Planet.

One stunning avenue of research is called artificial photosynthesis, which may one day give us an "artificial leaf," a more versatile form of photosynthesis that could make plants overall more efficient. We sometimes forget that photosynthesis is the end product of billions of years of totally random, chaotic chemical processes, and it developed these remarkable properties purely by chance. Hence, once quantum computers unravel the mystery of photosynthesis at the quantum level, we might be able to improve and modify the way plants grow. Billions of years of plant evolution might be squeezed into a few months on a quantum computer.

For example, Graham Fleming of the Kavli Energy NanoScience Institute at Berkeley says, "I really want to know how nature works in the early steps of photosynthesis. Then we could use that knowledge to create artificial systems that have all of the positive characteristics of the natural systems without all the baggage of having to produce seeds, maintain life, or defend themselves against bugs eating them."

Throughout history, plants were a mystery. They seemed to blossom by themselves, only occasionally needing water. Since ancient times, it was believed that plants grew by somehow eating the soil. It wasn't until the mid-1600s that this view changed. Jan van Helmont, a Belgian scientist, measured the weight of a plant and its soil. To his surprise, he found that the weight of the soil did not change at all over time. He concluded that plants grew because of the water.

Then the chemist Joseph Priestley conducted more detailed experiments, including one in which he put a plant in a glass jar along with a candle. He found that the candle burned out quickly if left alone, but could continue to burn in the presence of the plant, since the plant used up the carbon dioxide in the air and supplied oxygen for the candle.

By the early 1800s, biologists were beginning to fit together all the pieces, realizing that plants needed sunlight, water, and carbon dioxide, and would give off oxygen in the process.

Photosynthesis is so vital to the earth that it has literally reshaped the planet's atmosphere. When the earth was formed, its atmosphere in the early days was predominantly carbon dioxide, which came from the outgassing of ancient volcanoes. We see this in the atmospheres of Mars and Venus, which are made of almost pure carbon dioxide due to their volcanoes.

But when photosynthesis emerged on earth, it converted the carbon dioxide into the oxygen we now breathe. So with every breath, I am reminded of this momentous transition that occurred billions of years ago.

By the 1950s, scientists pieced together what is called the Calvin cycle, the complex chemical processes by which carbon dioxide and water turn into carbohydrates. Using various techniques including carbon-14 analysis, they could trace the movement of specific chemicals as they traveled through the plant.

Through these means biologists were able to slowly understand the life history of plants. But one step always eluded them. How do plants capture the energy of photons of light in the first place? What starts this long chain of events, beginning with the capture of the energy of sunlight? It remains a mystery to this day. But quantum computers may help unravel it.

Quantum Mechanics of Photosynthesis

Many scientists believe photosynthesis is a quantum process. It begins when photons, the discrete packets of light, hit a leaf that contains chlorophyll. This special molecule absorbs red and blue light, but not green, which is scattered back into the environment. Hence, the green color of plants is due to the fact that green is not absorbed by them. (If nature had created plants that absorb as much light as possible, plants would be black, rather than green.)

When light hits a leaf, you would expect it to be scattered in all directions and lost forever. But here is where quantum magic occurs. The photon of light impacts chlorophyll, and this creates energy vibrations on the leaf, called excitons, which somehow travel along the surface of the leaf. Eventually, these excitations enter what is called a collection center on the surface of the leaf, where the energy of the exciton is used to convert carbon dioxide into oxygen.

According to the Second Law of Thermodynamics, when energy is transformed from one form to another, much of that energy is lost into the environment. So one expects that much of the energy of the photon should dissipate when hitting the chlorophyll molecule and therefore become lost during this process as waste heat.

Instead, miraculously, the energy of the exciton is carried to the collection center with almost no energy loss at all. For reasons that are still not understood, this process is almost 100 percent efficient.

This phenomenon by which photons create excitons that pool in collection centers would be like a golf tournament where each golfer fires a ball randomly in all directions. Then, as if by magic, all these balls would somehow change direction and score a hole in one each time. This should not be happening, but it can actually be measured in the laboratory.

One theory is that this journey of the exciton is made possible by path integrals, which we saw earlier were introduced by Richard

Feynman. We recall that Feynman rewrote the laws of the quantum theory in terms of paths. When an electron moves from one point or another, it somehow sniffs out all possible paths between these two points. Then it calculates a probability for each route. Hence, the electron is somehow "aware" of all possible paths connecting these points. This means that the electron "chooses" the path with the most efficiency.

There is also a second mystery here. The process of photosynthesis happens at room temperature, where random motions of atoms in the environment should destroy any coherence among the excitons. Normally, quantum computers have to be cooled down to near absolute zero in order to minimize these chaotic motions, yet plants function perfectly well at normal temperatures. How is that possible?

Artificial Photosynthesis

One way to experimentally prove or disprove the existence of quantum effects is to look for indications of coherence, the telltale sign of quantum effects when atoms vibrate in unison. Normally, one would expect to find a chaotic jumble of individual vibrations, without any rhyme or reason, but if one detects some vibrations in phase with each other, this would immediately signal the presence of quantum effects.

In 2007, Graham Fleming reported that he saw this elusive phenomenon. He was able to announce the discovery of coherence in photosynthesis because he was using a special, ultrafast multidimensional spectroscope, which could generate pulses of light lasting a femtosecond (one millionth of one billionth of one second). He needed these exceptionally fast lasers to detect coherent light beams before random collisions with the environment destroyed the coherence. From the point of view of the laser, the atoms of the environment were almost frozen in time, and hence could be largely ignored. He was able to show that light waves could exist in two or more quantum states simultaneously. This meant that light could explore multiple pathways to the reaction center at the same time. This might explain why excitons could find the reaction center almost 100 percent of the time.

K. Birgitta Whaley, a colleague of Fleming's at Berkeley, adds, "The excitation effectively 'picks' the most efficient route . . . from a quantum menu of possible paths. This requires that all possible states of the traveling particle be superposed in a single, coherent quantum state for tenths of femtoseconds."

It might also explain how photosynthesis could operate at room temperature, without all the pipes and tubing found in a physics laboratory.

Quantum computers are ideally suited to making these quantum calculations. If this approach using path integrals is valid, then it means that we can now alter the dynamics of photosynthesis to solve a variety of problems. Instead of conducting thousands of experiments with plants, which takes an inordinate amount of time, these experiments could be done virtually.

For example, it may be possible to grow crops that are more efficient or produce more fruits and vegetables, increasing the yield of farmers.

Also, the human diet depends crucially on a handful of grains, such as rice and wheat, so a sudden blight that attacks our grains could upset the entire food chain. We would be helpless if just one of our basic foods were to be suddenly disrupted.

Scientists' new focus on the creation of an "artificial leaf" with artificial photosynthesis would help us be less dependent on this important natural process.

Artificial Leaf

When we discuss the world's biggest problems, CO_2 is usually described as one of the villains of the story. CO_2 captures energy from the sun and causes the earth to heat up. But what if we could recycle this greenhouse gas so it would become harmless? We might then also be able to create commercially valuable chemicals from recycled CO_2. Scientists propose that sunlight may be able to do exactly that. This new technology would take CO_2 from the air and combine it with sunlight and water to create fuel and other valuable chemicals, not unlike a leaf, but made artificially. Burning this fuel would create more CO_2, which could then recombine with sunlight and water to create more fuel, in a ceaseless process of recycling with no net gain of CO_2. In this way CO_2, which has been cast as the villain, becomes a useful resource.

For this recycling to work, it would proceed in two steps.

First, sunlight would be used to break apart water into hydrogen and oxygen. The hydrogen produced could then be used in fuel cells to power clean hydrogen cars. One problem with electric cars is that they use batteries, which, in turn, get their energy mainly from coal- and oil-fired plants. Although the electric battery burns cleanly, the electricity originally comes from polluting oil power plants. So there is currently a hidden cost to using electric batteries. Fuel cells, however, burn hydrogen and oxygen, which produces water as a waste product. So fuel cells burn cleanly, without oil and coal plants. However, the industrial infrastructure based on fuel cells is much less developed than for the electric battery.

Second, the hydrogen produced by splitting apart water can be combined with CO_2 to produce fuel and valuable hydrocarbons. This fuel, in turn, can be burned, and CO_2 is again produced, but it can be recombined with hydrogen and hence recycled. This could create a new cycle in which CO_2 could be continually reused so it doesn't build up in the atmosphere, stabilizing the amount of this greenhouse gas while providing energy at the same time.

"Our goal is to close the carbon fuel cycle," says Harry Atwater, director of the Joint Center for Artificial Photosynthesis (JCAP), a branch of the Department of Energy that funds artificial photosynthesis. "It's an audacious concept."

If successful, it would create a paradigm shift in the battle against global warming. CO_2 would be recast as just one cog in a larger wheel that keeps society moving. Quantum computers could play a decisive role in achieving carbon recycling. Writing in *Forbes* magazine, quantum researcher Ali El Kaafarani says, "quantum computers may be able to accelerate the discovery of new CO_2 catalysts that would ensure efficient carbon dioxide recycling whilst producing useful gases such as hydrogen and carbon monoxide."

Although this might sound like a dream, the first breakthrough took place in 1972, when Akira Fujishima and Kenichi Honda showed that light could be used to split water into hydrogen and oxygen, using one electrode made of titanium dioxide and another made of platinum. Although it was only .1 percent efficient, this proof-of-principle showed that it was possible to create an artificial leaf.

Since then, chemists have tried to modify this experiment to lower the cost, since platinum is very expensive. At JCAP, for example, chemists were able to use light to split apart water with an efficiency of 10 percent using an electrode made of a semiconductor and catalysts made of nickel.

The hard part is now to complete the final step and find a cheap way to combine hydrogen with CO_2 to create fuel. This is difficult because CO_2 is a remarkably stable molecule. Harvard chemist Daniel Nocera thinks he has found a viable way to accomplish this. He uses a bacterium, *Ralstonia eutropha,* which can combine hydrogen with CO_2 to create fuel and biomass, with an efficiency of 11 percent. Nocera says, "We did a complete artificial photosynthesis that's 10 to

100 times better than nature. . . . It's not a chemistry problem, necessarily, anymore. It's not even a technology problem." To him, the big problem has now been solved. Now, it's a question of economics, that is, whether industry and government will get behind recycling CO_2 given its cost.

Harvard's Pamela Silver, who works on this project, notes that using microbes to complete the carbon cycle may sound strange at first, but microbes are already used on an industrial scale to ferment sugar in the wine industry.

Meanwhile, Peidong Yang, a chemist at the University of California at Berkeley, also uses bioengineered bacteria, but in a different fashion. He uses light to split water into hydrogen and oxygen by using tiny semiconducting nanowires, and then grows bacteria on these nanowires, which then use the hydrogen to create various useful chemicals such as butanol and natural gas.

Quantum computers can take the technology to the next level. So far, much of the progress in this area is done by trial and error, requiring hundreds of experiments with exotic chemicals. For example, the process of using hydrogen to fix CO_2 into fuel is a complex molecular process, requiring the transfer of many electrons and the breaking of many bonds. Quantum computers may be able to replicate these chemical processes in a simulation and allow chemists to create new alternate quantum pathways. For example, CO_2 is the end product of a series of oxidation reactions. Quantum computers may be able to model ways in which to break the bonds of CO_2 so that they can recombine with hydrogen to create fuel.

If quantum computers provide the final step to creating artificial photosynthesis and the artificial leaf, it may open up entirely new industries that can provide new forms of efficient solar cells, alternate forms of crops, and new forms of photosynthesis. In the process, it might be possible to use quantum computers to find ways to recycle CO_2, which would go a long way in the effort to combat climate change.

So quantum computers may play a key role in harnessing the power of photosynthesis, which converts the energy of sunlight into food and nutrients. But in order to create a bounty of food, the next step is to have fertilizer to nourish the crops and help them flourish. Once again, quantum computers may play the decisive role in completing this last crucial step in order to feed the planet.

Ironically, the man who pioneered this last step, making it possible to feed billions of people and make modern civilization possible, is sometimes described not as one of the greatest scientists of all time, but as a war criminal.

CHAPTER 8

FEEDING THE PLANET

I n modern history, one man is responsible for saving more lives than any other person on earth, yet his name is largely unknown to the general public. It is reliably estimated that about half of humanity is alive today because of this man's discoveries, yet there are no biographies or documentaries singing his praises. Fritz Haber, a German chemist, touched the lives of every human on the planet. Haber was the man who discovered how to make artificial fertilizers. Fifty percent of all the food we eat is directly related to his pioneering research, yet his contribution is rarely celebrated by historians.

He unleashed the Green Revolution, breaking open nature's secrets to manufacture almost unlimited quantities of fertilizer that help feed the planet today. He changed world history when he discovered the crucial chemical process by which nitrogen could be taken from the air to create fertilizers. Where once peasants had to toil in the harsh soil to eke out a miserable living, today we have miles of green crops, as far as the eye can see. Instead of starving nations with barren, lifeless fields, we have lush farms yielding tremendous bounty.

But his role in history is tarnished by the fact that his stunning breakthrough can also be used to create devastating chemical weapons, including high-energy explosives as well as poison gas. Although billions of people on this planet owe their very existence to this man, his work also killed thousands who perished because of the havoc his discoveries unleashed on the battlefield.

Furthermore, we have to live with the fact that the Haber-Bosch process, as the technique he developed is known, is so power-hungry that it puts an enormous strain on the energy supply, exacerbating pollution and even climate change.

The problem, however, is that no one has been able to improve upon the Haber-Bosch process for a hundred years because it is so complicated at the molecular level. The hope, therefore, is that quantum computers will give us improved alternatives or modifications to Haber-Bosch so that we can feed the planet without soaking up so much energy and creating environmental problems.

But to appreciate the pioneering work of Haber, and the importance of quantum computers improving on his discoveries, one has to first appreciate his enormous contribution to escape the dismal destiny once predicted by Malthus.

Overpopulation and Famine

Back in 1798, Thomas Robert Malthus predicted that one day the population of the human race might exceed the food supply, resulting in mass starvation and death. To him, all animals were engaged in an eternal life-and-death struggle, and whenever their numbers exceeded the carrying capacity of their habitat, many would starve. Humans are no different. We too are bound by this iron law, that humanity can only flourish as long as there is enough food to eat. But since populations can grow exponentially, while food supply only marches forward slowly, the population might eventually outstrip the available food supply. This means there might be riots, mass starvation, followed by brutal warfare as nations fight for resources.

In the 1800s, it was increasingly apparent that this dreaded prophecy might come true. Although the human population was relatively stable at less than a million people for thousands of years, it was then experiencing a boom of unprecedented proportions. The coming of the Industrial Revolution and the Machine Age made possible a rapidly expanding population.

(I saw a graphic illustration of this when I was in grade school. In one experiment, we took a Petri dish full of nutrients and then put some bacteria in the middle of the dish. Within a few days, we saw that the bacteria expanded exponentially and created a large circular colony of cells, but then they suddenly stopped. Why had the bacteria stopped growing? I asked myself. And then I began to realize that the bacteria colony grew rapidly by consuming all the nutrients, and then died as a consequence of depleting the food supply. So this life-and-death struggle for food and growth was a Malthusian struggle in a Petri dish.)

Today, the world's food supply is heavily dependent on fertilizers. The essential ingredient of fertilizer is nitrogen, which is found in our protein and DNA molecules. Nitrogen, ironically, is the most plentiful chemical in the air we breathe, making up about 80 percent of it. For some mysterious reason, simple bacteria that can grow along the roots of legumes (e.g., in peanuts and beans) are able to extract nitrogen from the air and "fix" it with molecules of carbon, oxygen, and hydrogen to create ammonia, the essential ingredient needed to make fertilizer.

These bacteria have somehow mastered a puzzling chemical process. Although common bacteria can effortlessly extract nitrogen from the air to create life-giving fertilizers, chemists are still at a loss to duplicate Mother Nature so efficiently.

The reason is that the nitrogen we breathe in the air is actually N_2, i.e., two nitrogen atoms stuck together extremely tightly with three covalent chemical bonds. These bonds are so strong that normal chemical processes cannot break them. So chemists are saddled with this stubborn dilemma. The air we breathe is full of life-giving nitrogen, which in principle makes fertilizer possible, but it is of the wrong form, and hence useless.

It is like the proverbial man dying of thirst in an ocean full of salt water. You are surrounded by water but there's not a drop to drink.

We can easily see the problem by looking at the Schrödinger atom. Nitrogen has seven electrons, which can fill up the two available spaces in the 1S orbitals of the first energy level, and five electrons in the second level. To fill up all the orbitals of the first two levels requires ten electrons. (Recall that the electrons orbit in pairs, and that the first floor of the hotel has one room that holds two electrons, and the second has four rooms each holding two electrons.) This means that, at the second level, two electrons are in the 2S orbital, and the remaining three sit individually in the Px, Py, and Pz orbitals. So there are three electrons that are unpaired. When combined with a second nitrogen atom, this gives us three electrons shared between two atoms, reaching the ten electrons needed to fill the first two orbitals and, most importantly, giving us a triple bond, which is extremely strong.

Science for War and Peace

This is where the work of Fritz Haber comes in. Even as a child, he was fascinated by chemistry, often performing experiments by himself. His father was a prosperous merchant importing dyes and pigments, and he would sometimes help in his father's chemical factory. He was part of a rising generation of European Jews who were successful in business and science, but he eventually converted to Christianity. But above all, he was a nationalist, with a firm desire to help Germany with his knowledge of chemistry.

He focused on a number of chemical mysteries, including how to harness the nitrogen found in the air into useful products, such as fertilizer as well as explosives. He realized that the only way to split the two nitrogen atoms apart was to apply enormous pressure and temperature. By brute force, the nitrogen bonds could be broken, he theorized. He made history by finding the right magical combination in the laboratory. If you heated the nitrogen gas found in air to 300 degrees C and compressed it with the pressure of 200 to 300 times atmospheric pressure, then it was possible to finally break the nitrogen molecule apart and have it recombine with hydrogen to form ammonia, which is NH_3. For the first time in history, chemistry could be used to feed the world's rising population.

He would win the Nobel Prize in 1918 for this pioneering work. Today, about half the nitrogen molecules in your body are a direct consequence of Haber's discovery, so his enduring legacy is imprinted in your atoms. The world population today is over 8 billion people, and we could not feed this population without his work.

But his process is so energy hungry, requiring nitrogen to be compressed and heated to enormous pressures and temperatures, that it consumes 2 percent of the world's energy output.

Fertilizers were not the only thing on Haber's mind. Being a German nationalist, he was an enthusiastic supporter of the German army during World War I, and the energy stored in the nitrogen molecule could be harnessed to create life-giving fertilizer as well as fatal explosives. (Even amateur terrorists are aware of this process. A fertilizer bomb, capable of leveling an entire apartment building, consists of ordinary fertilizers saturated with fuel oil.) Thus, Haber used another by-product of his process, nitrates, to contribute to Germany's vast war machine, creating explosive chemical weapons in addition to poison gas that would take many innocent lives.

So, ironically, the man whose mastery of chemistry expanded the world population also doomed the lives of thousands of innocents. He is also known as the Father of Chemical Warfare.

But there is also a tragic aspect to his life. His wife, a pacifist, would commit suicide, perhaps due to her opposition to his research in chemical warfare and poison gas. Despite his decades of work supporting the government and the German military, he felt the wave of anti-Semitism sweeping the country in the 1930s. Although he was a Jew who had converted to Christianity, he left the country to seek refuge elsewhere and died of poor health in 1934. During World War II, the Nazi army would use Zyklon gas, a poison gas that was developed and perfected by Haber, to kill many of his own relatives in concentration camps.

ATP: Nature's Battery

Scientists who are anxious to apply quantum computers to the problem of replacing the inefficient Haber-Bosch process realize that they have to understand how nitrogen fixing is performed by Mother Nature.

In order to break the nitrogen bond, Haber's method was to apply high temperatures and enormous pressure from the outside. This is what makes it so inefficient. But nature does it at room temperature, without high-temperature furnaces and compressors. How can a lowly peanut plant do what usually takes a huge chemical plant?

In nature, the fundamental energy source is found in a molecule called ATP (adenosine triphosphate), which is the workhorse of life, nature's battery. Whenever you flex your muscles, take a breath, or digest food, you are using the energy from ATP to fuel your tissues. The ATP molecule is so elemental that it is found in almost all forms of life, indicating that it evolved billions of years ago. Without ATP, most of life on earth would die.

The key to understanding the secret of the ATP molecule is to analyze its structure. This molecule consists of three phosphate groups arranged in a chain, with each group consisting of a phosphorous atom surrounded by oxygen and hydrogen. The molecule's energy is stored in an electron located in the last phosphate group. When the body needs energy to perform its biological functions, it uses the energy stored in the electron in the last group.

When analyzing the nitrogen-fixing process in plants, chemists discovered that twelve molecules of ATP are required to supply the energy to break open a single N_2 molecule. Immediately, we can see the problem. Usually, atoms simply bump into each other one by one. If we have several atoms bumping into several more atoms, we see that this must take place in stages, because atoms bump into each other sequentially, not all at once. The process of ATP breaking down N_2 therefore goes through many, many intermediate steps.

In nature, harnessing energy from twelve ATP molecules from random collisions might take years. Clearly this is too slow to make life possible. So a series of shortcuts are necessary to greatly accelerate this process.

Quantum computers may be able to help solve this riddle. They could unravel this process at the molecular level, and perhaps improve the nitrogen-fixing process or find an alternative process.

As *CB Insights* magazine notes, "Using today's supercomputers to identify the best catalytic combinations to make ammonia would take centuries to solve. However, a powerful quantum computer could be used to much more efficiently analyze different catalyst combinations—another application of simulating chemical reactions—and help find a better way to create ammonia."

Catalysis: Nature's Shortcut

The key, scientists believe, is something called catalysis, which may be analyzed with quantum computers. A catalyst is like a bystander. It does not participate directly in a chemical process, but somehow by its presence it facilitates a reaction.

Normally, chemical reactions found in the body are quite slow, sometimes taking place over long periods of time. Sometimes, something magical happens to speed up these processes so they can take place in a fraction of a second. This is where catalysts come in. For the nitrogen-fixing process, there is a catalyst called nitrogenase. Like a conductor, its purpose is to orchestrate the many steps necessary to combine twelve ATP molecules with nitrogen to break the triple bond. So nitrogenase is the key to creating a Second Green Revolution. But unfortunately our digital computers are too primitive to unravel its secrets. A quantum computer, however, may be perfectly suited for this important task.

Catalysts like nitrogenase work in two stages. First, they bring two reactants together. The catalyst and the reactants fit together like a jigsaw puzzle, allowing the two reactants to bind. Second, the energy required for a reaction to occur, called the activation energy, is sometimes too high for the reactants to interact with each other. The catalyst, however, lowers the activation energy so that the reaction can proceed. Then the reactants can combine to create a new chemical and leave the catalyst intact.

To understand how a catalyst works, think of a matchmaker, trying to bring a potential couple together who might live in two different cities. Normally, the chances of a purely random meeting between these two is extremely small, since they move in entirely different circles many miles apart. But a matchmaker can make contact with both parties and bring them together, vastly increasing the chances that something will happen between them. Almost all important chemical processes in the body are mediated by some catalyst.

Now, let us introduce a quantum matchmaker, who realizes that sometimes you have to nudge the couple to get them to bond with each other. For example, perhaps one person is shy, reticent, or nervous. Something prevents them from breaking the ice. In other words, they have to overcome an activation barrier before they can start their relationship. That's what the quantum matchmaker does, breaking the ice or helping them pass through the barrier that separates them. This is called tunneling, a bizarre feature of the quantum theory in which one can penetrate seemingly impenetrable barriers. Tunneling is the reason why radioactive elements like uranium can emit radiation, because the radiation tunnels its way through a nuclear barrier to reach the outside world. The process of radioactive decay, which heats up the center of the earth and drives continental drift, is due to tunneling. So the next time you see a gigantic volcano blow its top, you are seeing the power of quantum tunneling. By the same token, ATP molecules may be able to magically "tunnel" through this energy barrier and complete the chemical reaction.

We will see, moreover, that nearly all the key reactions that make life possible require catalysts, and the origin of life itself might be due to quantum mechanics.

Sadly, nitrogenase and nitrogen fixing are so complex that progress, though steady, has been slow. Although scientists now have a complete molecular diagram of what the nitrogenase molecule looks like, it is so complicated that no one knows precisely how it works. This entire process is so fraught that it is hopeless for a digital computer to unravel its secrets. This is where quantum computers can excel, by filling in all the steps that make this possible.

One company investigating this ambitious project is Microsoft. On the heels of its success with commercial ventures like the Xbox, it has been looking into projects that are riskier but potentially lucrative. Even as far back as 2005, Microsoft was interested in blue-sky projects like quantum computers. Back then, Microsoft set up a company called Station Q to investigate problems like nitrogen fixing and quantum computation.

"I think we're at an inflection point in which we are ready to go from research to development," says Todd Holmdahl, corporate vice president of Microsoft's quantum program. "You have to take some amount of risk in order to make a big impact in the world, and I think we're at the point now that we have the opportunity to do that."

He likes to compare this to the invention of the transistor. Back then, physicists were scratching their heads trying to think of practical applications for their invention. Some thought the transistor was just useful for signaling ships at sea. Likewise, the creation of Microsoft's quantum computer, which *The New York Times* compared to "science fiction," may also transform society in unexpected ways.

Microsoft is one company that cannot wait to solve the nitrogen-fixing problem. It already is using first-generation quantum computers to see if the mystery of this process can be uncovered. The implications are profound, with the potential to create a Second Green Revolution and feed an exploding world population with lower energy costs. Failure to do so could have disastrous side effects, as we've seen, perhaps leading to riots, famine, and wars.

Recently, Microsoft had a setback when some experimental results on topological qubits did not turn out correctly, but for the true believers in quantum computers, that is just a speed bump.

In fact, Google's CEO, Sundar Pichai, recently claimed that he thinks that quantum computers may be able to improve on the Haber process within a decade.

Quantum computers will be essential to analyzing this important chemical process, in several ways:

- Quantum computers can help elucidate this complex process, atom for atom, by solving the wave equation for the various components within nitrogenase. This will help illuminate all the many missing steps in nitrogen fixing.
- They may virtually test different ways to break the N_2 bond, other than by brute force or by catalysis.
- They can model what would happen if we replaced various atoms and proteins with substitutes, to see if one can make the process of nitrogen fixing more efficient, less energy intensive, and less pollutive, with different chemicals.
- Quantum computers can test various new catalysts to see if they can speed up the process.
- Quantum computers may test different versions of nitrogenase, with different arrangements of protein chains, to see if one can improve on its catalytic properties.

So if Microsoft and others can solve the mystery of nitrogen fixing, it could have an enormous impact on our food supply. But scientists have other dreams for quantum computers. They want not just to solve the problem of energy-efficient food production, they want to understand the nature of energy itself. Can quantum computers solve the energy crisis?

CHAPTER 9

ENERGIZING THE WORLD

At first glance, one might suspect that the titans of twentieth-century industry, Thomas Edison and Henry Ford, would be bitter rivals. After all, Edison was the tireless, driving force behind the electrification of industry and society. With 1,093 patents, he revolutionized our way of life with numerous inventions we now take for granted, powered by electricity. Whereas Ford, however, made his millions from the Model T, which was powered by fossil fuels. He helped create the modern industrial infrastructure built on oil. To him, burning oil and gasoline would power the future.

In reality, Edison and Ford were close friends. In fact, as a young man, Ford idolized Edison. For years, they would vacation together and delight in each other's company. Perhaps they became quite close because they both created world-class companies through sheer force of will.

Edison and Ford would pass the time by making wagers, betting on which energy source would power the future. Edison favored the electric battery, while Ford believed in gasoline. For anyone listening to this wager, it was a no-brainer. One would surely conclude that Edison would win handily. Electric batteries were quiet and safe. Oil, by contrast, was noisy, noxious, and even dangerous. The idea of having a gas station every few blocks was considered preposterous.

In many ways, the critics of oil were all correct. The fumes emitted by the internal combustion engine can cause respiratory illnesses and accelerate global warming, and gasoline-powered cars are still noisy.

But it was Ford who eventually won the bet.

Why is that?

For one thing, the energy packed in a battery is a tiny fraction of the energy in a gallon of gasoline. (The best batteries can store about 200 watt-hours per kilogram of energy, while gasoline can store 12,000.)

And when huge oil fields were discovered in the Middle East, Texas, and elsewhere, the price of gasoline plummeted, putting the automobile within the reach of working-class Americans.

People began to forget about Edison's dream. Inefficient, clumsy, and weak, the electric battery could not compete with cheap, high-octane fuel designed for an energy-hungry population.

Because Moore's law has revolutionized the world economy with cheap computer power, there is a tendency to assume that everything obeys this law. We are puzzled, therefore, by the fact that battery power efficiency has lagged for so many decades. We forget that Moore's law only applies to computer chips, and that chemical reactions like the ones that power batteries are notoriously hard to predict. Forecasting new chemical reactions that would increase the efficiency of a battery is a major undertaking.

Instead of tediously testing hundreds of different chemicals for their performance in a battery, in the future it will be much quicker and cheaper to simulate their performance with a quantum computer. Like the simulations that may help unravel the secrets of photosynthesis or natural nitrogen fixing, "virtual chemistry" may one day replace the arduous trial and error in chemistry laboratories.

Solar Revolution?

This challenge of increasing battery performance has tremendous economic implications. Back in the 1950s, futurists proclaimed that our houses would one day be powered by sunlight. Vast arrays of solar cells, supplemented with powerful windmills, would capture the energy of the sun and the wind and provide cheap and reliable energy. Energy for free. That was the dream.

However, reality turned out differently. Renewable energy has dropped in cost over the decades, but at an agonizingly slow rate. The arrival of a Solar Age has been slower than people anticipated.

In part, the problem lies in the limitations of modern batteries. When the sun does not shine and the winds do not blow, power from renewable energy drops to zero. The weak link in the chain of renewable energy is storage—how you store energy for a rainy day. While computer speed grows exponentially as we systematically miniaturize silicon chips, battery power only grows when we discover new efficiencies or even new chemical compounds. Currently, it still uses chemical reactions that were known in the last century. If a super battery could be built with increased efficiency and power, it could greatly accelerate the transition to a carbon-free energy future and blunt global warming.

History of the Battery

Looking back, we see that the history of the battery has moved at a glacial pace across the centuries. In ancient times, it was well known that if one walked across a carpet, you might get an electric shock when touching a doorknob. But this was just a curiosity, until history was made in 1786, when physicist Luigi Galvani was rubbing a piece of metal against the severed legs of a frog. He noticed, much to his surprise, that the legs twitched by themselves.

This was a pivotal discovery, because scientists could now show that electricity could drive the movement of our muscles. In one instant, scientists realized that you did not have to appeal to some mythical "life force" to explain how inanimate objects could become animate. Electricity was the key to understanding how our bodies could move without spirits. But these pathbreaking studies in electricity also inspired one of his intrepid colleagues.

In 1799, Alessandro Volta built the first battery and showed that he could create a chemical reaction to reproduce this effect. To create electricity in the laboratory on demand was a sensational discovery. News spread quickly that this strange force could now be generated at will.

But sadly, the battery has not changed much in over 200 years. The simplest battery starts with two metal rods or electrodes placed in separate cups. In both cups is a chemical called an electrolyte, which allows a chemical reaction to take place. Connecting the two cups is a tube in which ions can pass from one cup to the other.

Because of the chemical reaction in the electrolyte, electrons leave one electrode, called the anode, and pass onto the other electrode, called the cathode. The movement of electric charges has to be balanced, so while negatively charged electrons pass from the anode to the cathode, there is also a movement of positive ions through the electrolyte-connecting tube. The flow of these charges creates electricity.

This basic design has not changed in several centuries. What has changed is mainly the chemical composition of the various components. Chemists tediously experimented with different metals and electrolytes in order to maximize the electric voltage or increase its energy content.

Because it was widely believed that there was no market for an electric car, there was little pressure to improve the technology.

Lithium Revolution

In the postwar era, battery technology was a backwater field. Progress stagnated because there was relatively little demand for electric vehicles and portable electronic appliances. However, increased concern about global warming and the exploding electronics market has sparked new research in battery technology.

Because of the threat of pollution and global warming, the public has demanded action. As pressure mounted on the automobile industry to convert to electric cars, inventors rushed to create more powerful batteries. Batteries were gradually becoming competitive with gasoline.

One success story has been the introduction of the lithium-ion battery, which has taken the market by storm. They are found in nearly all forms of electronics, in cell phones, computers, and even jumbo jetliners. What makes them so ubiquitous is the fact that they have the highest energy capacity of any battery available, yet they are portable, compact, reliable, and efficient. It is the end product of decades of research, painfully analyzing hundreds of different chemicals for their electrical properties.

What makes them so convenient is the nature of the lithium atom. When we look at the periodic table of elements, we see that it is the lightest of all the metals, which is important when we want lightweight batteries for cars and planes.

We also see that it has three electrons orbiting the nucleus. The first two electrons fill up the lowest energy level of the atom, the 1S shell, so the third electron, in a higher orbit, is loosely bound, making it easy to remove and energize the battery. This is one reason why it is so easy to generate an electric current with the lithium battery.

Putting it all together, the lithium-ion battery has an anode made of graphite, a cathode made of lithium cobalt oxide, and an electrolyte made of ether. The impact of lithium-ion batteries has been so revolutionary that the Nobel Prize in Chemistry was given to several scientists who perfected it: John B. Goodenough, M. Stanley Whittingham, and Akira Yoshino.

However, one undesirable feature of lithium-ion batteries is that, although they have the highest energy density of any battery on the market, they still have only 1 percent of the energy stored in gasoline. If we are to enter a carbon-free era, we need a battery with an energy density approaching that of its fossil fuel rival.

Beyond Lithium-Ion Batteries

Because of the enormous commercial success of the lithium-ion battery, found everywhere in modern society, there is a feverish search under way for a replacement or improvement for the next generation. Again, engineers are limited by their trial-and-error approach.

One such candidate is the lithium-air battery. Unlike other batteries, which are completely sealed, this one allows air to flow in. The oxygen from the air interacts with the lithium, releasing the battery's electrons (and creating lithium peroxide).

The big advantage of the lithium-air battery is that its energy density is ten times that of the lithium-ion battery, so it is approaching the energy density of gasoline. (This is because the oxygen comes for free from the air, rather than having to be stored within the battery itself.)

Despite the enormous boost in energy density found in lithium-air batteries, a host of technical problems have prevented this remarkable battery from working in practice. In particular, it has a short life span of only two months or so. Scientists who have faith in this technology believe that, by experimenting with scores of different types of chemicals, one might be able to solve many of these technical problems.

In 2022, the Japanese National Institute for Materials Science, working with the investment company SoftBank, announced a promising new type of lithium-air battery that has a much higher energy density than the standard lithium-ion battery. However, details are still not available to see if they have surmounted the series of problems facing this promising technology.

One persistent annoyance to owning an electric car is the time it takes to charge a battery, which can run from several hours to a day. So another technology being pursued is the SuperBattery, a hybrid system created by Skeleton Technologies and the Karlsruhe Institute of Technology in Germany, which promises to be able to charge an electric vehicle in as little as fifteen seconds.

On the one hand, it uses a standard lithium-ion battery. But what makes it novel is that the SuperBattery combines the lithium-ion battery with a capacitor to reduce the charging time. (A capacitor stores static electricity. At its simplest, it consists of just two parallel plates, one charged positively and the other charged negatively. The great advantage of capacitors is that they can store electrical energy and then release it very rapidly.) The use of supercapacitors to offer rapid charging has also attracted other firms as well.

Because the potential rewards are huge, a number of enterprising groups are hard at work on the successor to lithium-ion batteries. These include the following experimental technologies:

- NAWA Technologies claims that its Ultra Fast Carbon Electrode, using nanotechnology, can boost battery power ten times and increase its life span five times. It claims that an electric car's range could become 1,000 kilometers, with a charging time of only five minutes to reach 80 percent capacity.
- Scientists at the University of Texas claim that they are able to remove one of the least desirable components from its batteries, cobalt. Cobalt is costly and toxic, and they claim to be able to replace it with manganese and aluminum.
- The Chinese battery cell manufacturer SVOLT has announced that they too can replace cobalt in their

batteries. They claim to be able to increase the range of electric vehicles to 500 miles and improve the battery life.

- Scientists at the University of Eastern Finland have developed a lithium-ion battery with a hybrid anode, using silicon and carbon nanotubes, which they claim increases the performance of the battery.

- Another group looking into silicon are the scientists at the University of California at Riverside. They use the basic lithium-ion battery, except they replace the graphite anode with silicon.

- Scientists at Monash University in Australia have replaced the lithium-ion battery with a lithium-sulfur battery. They claim that their battery can power a smartphone for five days or an electric vehicle for 620 miles.

- IBM Research and others are looking into replacing toxic elements like cobalt and nickel and even the lithium-ion battery itself with seawater. IBM claims that a seawater battery would be cheaper and have a higher energy density.

While incremental improvements are being made to the lithium-ion battery, the basic strategy introduced 200 years ago by Volta is still with us. The hope is that quantum computers may enable scientists to systematize this process, making it cheaper and more efficient, so that millions of experiments may be conducted virtually.

The problem is that the complex chemical reactions found inside a battery do not obey any simple law, like Newton's mechanics. But quantum computers may be able to do this heavy lifting, simulating complicated chemical reactions without actually performing them.

Not surprisingly, the automotive industry is investing in quantum computers to see if a super battery might be designed using pure mathematics. A super-efficient battery could remove the chief bottleneck preventing a Solar Age: the storage of electricity.

The Automotive Industry and Quantum Computers

One company that sees the potential of quantum computers to revolutionize their industry is the automotive giant, Daimler, which owns Mercedes-Benz. As early as 2015, Daimler created the Quantum Computing Initiative to keep abreast of this rapidly changing field.

Ben Boeser, who is part of their Mercedes-Benz Research and Development North America group, says, "It's a very research-oriented activity, looking at things that happen 10 to 15 years out, but we want to understand the basics as a new universe is created—and we as a company want to be part of it." Daimler sees quantum computing not just as a scientific curiosity, but as part of their bottom line.

Holger Mohn, editor of Daimler's online magazine, points out the other benefits of quantum computing besides finding new battery designs. He writes, "It could become the best way to discover new, more efficient technology, simulating aerodynamic shapes for better fuel efficiency and a smoother ride, or to optimize manufacturing processes with myriad variables." In 2018, Daimler assembled a network of top engineers to work closely with both Google and IBM to develop the technology necessary to crack some of these vexing problems. Already, they are writing codes and uploading them to the cloud to familiarize themselves with quantum computing.

For example, the basic equations of aerodynamics are well known. But instead of running costly wind tunnel tests to reduce air friction on their cars, it is much cheaper and more convenient to put their cars in a "virtual wind tunnel," i.e., test the efficiency of their car design in the memory of a quantum computer. This will allow rapid analysis in order to reduce drag.

Airbus is using a quantum computer to create a virtual wind tunnel to calculate the most fuel-efficient path in which its airplanes ascend and descend. And Volkswagen is also using this technology to calculate the optimal path for buses and taxis to take in a congested city.

Since 2018, BMW has looked into quantum computers to solve a host of problems, using Honeywell's latest quantum computer. Several avenues they are investigating include:

- Creating a better car battery
- Determining the best places to install electrical charging stations
- Finding more efficient ways to purchase the variety of components that go into BMW cars
- Increasing aerodynamic performance and safety

In particular, BMW is looking to quantum computers to help with optimization programs, i.e., lowering costs while increasing performance.

But quantum computers are not just useful for creating newer, cheaper, and more powerful batteries and cars, without destroying the environment. Quantum computers may eventually also free us from the dangers of dreaded, incurable diseases that have afflicted humanity since the dawn of time. We now turn to how quantum computers can create a revolution in medicine.

The Fountain of Youth, instead of being a fabled wellspring of eternal life, might turn out to be a quantum computer.

PART III

QUANTUM MEDICINE

CHAPTER 10

QUANTUM HEALTH

H ow long can you live?

For most of human history, the average life expectancy for humans hovered between twenty and thirty years. Life was often short and miserable. People lived in constant fear of the next plague or famine.

Stories from the Bible and other ancient texts are full of tales of pestilence and disease. Later, these stories were full of orphans and evil stepmothers because parents often did not live long enough to raise their own children.

Sadly, throughout history, physicians were little more than quacks and charlatans, pompously dispensing "cures" that often made the patient worse. The rich could afford private doctors, who jealously guarded their useless potions, while the poor often died in poverty in filthy, overcrowded hospitals. (All this was parodied by the French playwright Molière in the hilarious play *Le Médicin Malgré Lui*, or *The Doctor in Spite of Himself*, in which a poor peasant is mistaken for a prominent physician, who then deceives everyone by using big, fancy, made-up Latin words to offer silly medical advice.)

However, several historic advances took place that lengthened our life expectancy. First was the arrival of better sanitation. Ancient cities used to be cesspools of rotten food and human waste. People would regularly throw their garbage out on the street. The roads of ancient cities often resembled a smelly obstacle course, a breeding ground for diseases. But in the 1800s, citizens denounced these unsanitary conditions, leading to the creation of a sewage system and improved sanitation, which eliminated scores of deadly waterborne diseases, adding perhaps fifteen to twenty years to our life expectancy.

The next revolution took place because of the bloody European wars that engulfed the continent during the 1800s. There were so many soldiers dying of gaping battle wounds that kings and monarchs would decree that cures that really worked would be royally rewarded. Suddenly ambitious doctors, instead of just trying to impress rich patrons with useless concoctions, were publishing articles about therapies that actually helped patients. Medical journals began to flourish, documenting advances based on experimental proof, not just the reputation of the author.

With this new orientation among doctors and scientists, the stage was set for revolutionary advances like antibiotics and vaccines, which would eventually vanquish a bestiary of deadly diseases, adding perhaps ten to fifteen more years to the average life expectancy. Better nutrition, surgery, the Industrial Revolution, and other factors also contributed to the increase in life expectancy.

So now the average life expectancy in many countries is in the seventies.

Unfortunately many of these breakthroughs in modern medicine were due to luck, not careful design. There was nothing systematic about finding cures for these diseases, which occurred mainly through fortuitous accidents.

For example, in 1928, when Alexander Fleming inadvertently observed that particles of bread mold could kill bacteria growing in a Petri dish, he set off a revolution in health care. Doctors, instead of helplessly watching their patients die of common diseases, could now give antibiotics like penicillin, which, for the first time in human history, could actually cure the patient. Soon, there were antibiotics against cholera, tetanus, typhoid, tuberculosis, and a host of other diseases. But most of these cures were found by trial and error.

Rise of Drug-Resistant Germs

Antibiotics have been so effective and so often prescribed that now the germs are fighting back. This is not an academic question, because drug-resistant germs are one of the major health issues facing society today. Deadly diseases that were once banished, like tuberculosis, are now slowly coming back in virulent, incurable form. These "super-bugs" are often immune to the latest antibiotics, leaving the general population helpless against them.

Furthermore, as humanity expands into previously unexplored and unpopulated areas, we are constantly exposed to new diseases for which we have no immunity. So there is a huge pool of unknown diseases waiting to jump out and infect humanity.

Some believe that the large-scale use of antibiotics in animals has accelerated this trend. Cows, for example, become breeding grounds for drug-resistant germs because farmers sometimes overadminister antibiotics in order to increase milk and food production.

Because of the threat that these diseases may come back stronger than ever, there is an urgent need for a new generation of antibiotics that are cheap enough to justify their cost. Sadly, there has been no development of new classes of antibiotics for the last thirty years. The antibiotics that our parents used are about the same ones we use today. One problem is that thousands of chemicals must be tried in order to isolate a handful of promising drugs. It costs about $2 to $3 billion to develop a new class of antibiotics by these methods.

How Antibiotics Work

Using modern technology, scientists have gradually deduced how certain kinds of antibiotics work. Penicillin and vancomycin, for example, interfere with the production of a molecule called peptido-glycan, which is essential for creating and strengthening the cell wall of the bacteria. These drugs therefore cause the bacteria's walls to fall apart.

Another class of drugs, which are called quinolones, throws a monkey wrench into the bacteria's reproductive chemistry, so that its DNA does not function properly and hence cannot reproduce.

Another, which includes tetracycline, interferes with the bacteria's ability to synthesize a key protein. And yet another class stops the cells from producing folic acid, which in turn interferes with the bacteria's ability to control chemicals flowing across the cell wall.

Given these advances, why is there a bottleneck?

For one, these new antibiotics take a long time to develop, often over ten years. These drugs must be carefully tested to make sure that they are safe, which is a time-consuming and costly process. And after a decade of hard work, the final product often cannot pay the bills. The bottom line for many pharmaceutical companies is that the sales must compensate for the cost to make these drugs.

Role of Quantum Medicine

The problem is that, like battery designs since the time of Volta, the basic strategy hasn't changed much since Fleming's age. Basically, we still blindly test various candidates against germs inside a Petri dish. Today, using automation, robotics, and mechanized assembly lines, thousands of Petri dishes containing different types of pathogens can be exposed to promising drugs all at once, mimicking the basic approach pioneered by Fleming 100 years ago.

Since then, our strategy has been:

Test promising substance → determine if it kills bacteria → identify the mechanism

Quantum computers might upend this process entirely and accelerate the search for new lifesaving drugs. They are powerful enough that one day they might guide us systematically to new ways to destroy bacteria. Instead of spending ages fiddling with different drugs over many decades, we might be able to rapidly design new drugs inside the memory of a quantum computer.

This means reversing the order of the strategy:

Identify the mechanism → determine if it kills bacteria → test promising substance

If, for example, the basic mechanism by which these antibiotics can kill germs is unraveled at the molecular level, one might be able to use that knowledge to create new drugs. This means that first, you start with the mechanism you desire such as breaking down the cell wall of the bacteria, then use quantum computers to determine how to do this by finding weak spots in a bacteria's wall. Next, you test different drugs that can carry out this function and finally focus on the handful that actually work against the bacteria.

For example, trying to model the penicillin molecule with a conventional computer faces an enormous challenge. Doing so would require 10^{86} bits of computer memory, far beyond the capability of any digital computer. But this is within the capability of a quantum computer. So trying to discover new drugs by analyzing their molecular behavior can be a prime target for quantum computers.

Killer Viruses

Similarly, modern science has been able to attack viruses using vaccines, but only up to a point. Vaccines work indirectly by stimulating the immune system of the body, rather than by directly attacking the virus, so progress to cure diseases caused by viruses has been slow.

One of the greatest killers in history is smallpox, which has killed

300 million people since 1910 alone. Smallpox was known even in ancient times. It was also known that if someone had the disease and recovered, then their scabs could be made into a powder and given to a healthy person via breaks in their skin. That person would be inoculated against the disease.

In 1796, this technique was refined and successfully used in England. Physician Edward Jenner took pus from milkmaids who recovered from cowpox, which resembles smallpox. He then injected the pus into healthy individuals, who developed immunity against smallpox.

Since then, vaccines have been used against a large number of previously incurable diseases, such as polio, hepatitis B, measles, meningitis, mumps, tetanus, yellow fever, and many others. There are thousands of possible vaccines that might have therapeutic value, but without an understanding of how the body's immune system works at the most minute scale, it is impossible to test all of them.

Instead of testing each vaccine experimentally, one might be able to "test" them within a quantum computer. The beauty of this method is that the search for new vaccines can take place rapidly, cheaply, and efficiently, without using messy, time-consuming, and expensive trials.

In the next chapter, we will explore how quantum computers might be able to modify and strengthen our immune system, protecting us against cancer and perhaps currently incurable diseases like Alzheimer's and Parkinson's. But first, there's one more way quantum computers can help defend us against the next global pandemic virus.

Covid Pandemic

One way to see the power of quantum computers is to consider the tragedy of the Covid pandemic, which has killed about a million people in the U.S. so far and plunged billions of people around the world into economic hardship and distress. Quantum computers, however, can give us an early warning system to detect emerging viruses before they spawn a worldwide pandemic.

Sixty percent of all diseases, it is believed, originally come from the animal kingdom. So there is a vast reservoir of new germs that can generate a host of new diseases. And as human civilization expands into previously undeveloped areas, we are exposed to new animals and their illnesses.

For example, using genetic analysis, one can determine that the flu virus mainly originated in birds. Many flu viruses emerge in Asia, where farmers engage in something called polyfarming, which involves living in close proximity to pigs and birds. While the virus originates in birds, pigs often eat the bird droppings, and humans eat the pigs. So the pigs act like a mixing bowl, combining the DNA of birds and pigs to create new viruses.

Similarly, the AIDS virus has been traced back to simian immunodeficiency virus (SIV), which infects primates. Using genetics, scientists have conjectured that someone in Africa ate the flesh of a primate sometime between 1884 and 1924, which then mixed with the DNA of a human to create HIV, a mutated version of SIV that can attack people.

With advances in transportation, increased travel across the globe accelerated the spread of diseases like the plague during the Middle Ages. Historians have tracked the paths taken by ancient mariners as they went from city to city, thereby spreading the plague to distant shores. By comparing when the ships docked at a certain port to the date of the outbreak of the disease, one can see how the plague spread across the Middle East and Asia, jumping from city to city. Today, we have jetliners that can spread a disease across continents within a matter of hours.

So it is only a matter of time before another pandemic, spread by international jet travel, grips the world.

But because of remarkable advances in genomics, in 2020 scientists were able to sequence the genetic material of the Covid-19 virus within just a few weeks. This allowed scientists to create vaccines that would stimulate the body's immune system to attack the virus. But this was just tweaking the body's own immune system to be able to defend itself. What was missing was a systematic way to defeat this deadly virus itself.

Early Warning System

There are several ways in which quantum computers can help stop the next pandemic. At the very least, we need an early warning system to detect the virus as it emerges in real time. From the moment a new version of the Covid-19 virus emerges, it takes weeks before an alert can be issued. During that period the virus can escape unnoticed into the human ecosystem. A delay of a few weeks can allow the virus to spread to millions of people.

One method for tracking epidemics is to put sensors in sewer systems around the world. Viruses can easily be identified by analyzing the sewage, especially around crowded urban areas. Rapid antigen tests can spot the outbreak of a virus within about fifteen minutes. However, the data emerging from millions of sewer systems can easily overwhelm digital computers. But quantum computers excel at analyzing mountains of data to find that missing needle in the haystack. Already, certain communities around the country are inserting sensors in their sewer system as an early warning system.

Another early warning system was demonstrated by the Kinsa company, which makes thermometers that are connected to the internet. By examining the fevers erupting across the country, one can detect important anomalies. For example, in March 2020, hospitals in the American South were flooded with strange reports of thousands of people suffering from a new virus. Many died. Hospitals were overwhelmed.

One theory is that the Mardi Gras celebration in late February

2020 in New Orleans was a superspreader event that exposed hundreds of thousands of unsuspecting people to the virus. Sure enough, when analyzing the thermometer readings right after Mardi Gras, one can see a sudden spike in patient temperatures in the South. Sadly, because doctors had no experience dealing with the new deadly virus, it took weeks after Mardi Gras to alert doctors to the pandemic. Many died because of this critical delay in identifying the virus, the appearance of which caught the medical establishment totally by surprise.

In the future, with a vast network of medical devices like thermometers and sensors connected to the internet, one might have an instantaneous temperature readout of what is happening around the country analyzed by quantum computers. With a simple glance at the map of the country, one can see hotspots representing a potential new superspreader event.

Another way to create an early warning system is to utilize social media, which better than anything else gives us the pulse of what is happening around the country in real time. For example, algorithms of the future would be primed to look for anomalous postings on the internet. If, for example, people begin to say things like "I can't breathe" or "I can't smell," these anomalous phrases could be picked up by quantum computers. Then health care workers can follow up on these incidents to see if they are being caused by a transmissible disease.

Similarly, quantum computers may be able to detect outbreaks of the virus as they happen. Sensors may be developed that can detect aerosols of the virus floating in the air. At the beginning of the epidemic, government officials asserted that staying six feet away from others was sufficient to prevent the spread of the virus. Transmission, they claimed, occurred mainly through large droplets due to coughing and sneezing.

It is now believed that this was probably incorrect. Actual studies of the virus show that aerosol particles after a sneeze, for example, can carry the virus twenty feet or more. In fact, it is now thought that one of the main ways in which the virus spreads is through aerosols generated simply by talking. Sitting next to people who are singing, chanting, and speaking loudly indoors for more than fifteen minutes is one way to accelerate the spread of the virus.

So in the future, a network of sensors placed indoors might be able to detect aerosols in the air and then send the results to quantum computers, which can analyze this vast pool of information to find the early warning signs of the next pandemic.

Deciphering the Immune System

Vaccines have proven that the body's own immune system is a powerful defense against infectious disease. But scientists know very little about how it actually works.

We are still learning surprising new things about the immune system. For example, scientists now realize that many diseases do not directly attack the body. The Spanish flu of 1918 killed more people than all who died in World War I. Unfortunately, samples of the virus were not preserved, so it is difficult to analyze the virus and determine how it killed people. But several years ago, scientists were able to visit the Arctic and examine the bodies of those who died of the virus but were preserved in the permafrost.

What they found was interesting. The disease did not directly kill its victim. What it did was overstimulate the body's own immune system, which then began to flood the body with dangerous chemicals in the hope of killing the virus. This cytokine storm is what eventually killed the patient. So the main killer was actually the body's own immune system gone berserk.

A similar story was found with Covid-19. When people are committed to the hospital, their situation at first may not seem dire. But in the late stages of the disease, when the cytokine storm sets in, the dangerous chemicals that flood the body eventually cause organs to fail. Death often results if this is untreated.

In the future, quantum computers may provide an unprecedented look into the molecular biology of the immune system. This may present numerous ways in which to turn off or dial down the immune system so it does not kill you in the event of a serious infection. We will discuss the immune system in more detail in the next chapter.

Omicron Virus

Quantum computers may also prove critical in determining the properties of a virus as they are mutating. For example, the Omicron variant of Covid-19 emerged around November 2021. Its genome was sequenced, and alarm bells went off immediately. It had fifty mutations, which made it more transmissible than the Delta virus. But scientists were helpless to determine precisely how dangerous these mutations would make it. Do they allow the spike proteins to enter into human cells much faster than before, and hence wreak havoc on the human race? They could only wait and see. In the future, quantum computers might be able to determine how lethal a virus is by analyzing the mutations in its spike proteins, rather than waiting for weeks with our fingers crossed.

One may be able to predict the course of this and other viruses as soon as we know their structure. Digital computers today are too primitive to simulate how a virus like Omicron can attack the human body. But once we know the precise molecular structure of the virus, we might be able to use quantum computers to simulate the specific effects of the virus on the body, so we know ahead of time how dangerous it is and how to combat it.

Fortunately, we also have evolution on our side. Many ancient diseases that killed a large portion of the human race, like the Spanish flu virus of 1918, are probably still with us, but in mutated form as an endemic rather than pandemic. According to evolutionary theory, different strains of a virus are in competition with each other. Thus, there is an evolutionary pressure to become more infectious and outrace the competition. So each generation of mutations might be more infectious than the previous one. But if you kill off too many people, then you don't have enough hosts to continue spreading. Hence, there is also evolutionary pressure to be less lethal.

So, in other words, in order to remain in circulation many viruses evolve so they become more infectious, but less lethal. So perhaps we will just have to learn to live with the Covid virus, except in less lethal form.

The Future

Antibiotics and vaccines are the foundation of modern medicine. But antibiotics are usually found by trial and error, and vaccines only stimulate the immune system to create antibodies to fight off a virus. So one of the goals of modern medicine is to develop new antibiotics, and another is to understand the body's immune response, which is our first line of defense against viruses and also one of the greatest killers of all time, cancer. If the mystery surrounding our immune system can be solved using quantum computers, then we will also have a way to attack some of the greatest incurable diseases, such as certain forms of cancer, Alzheimer's, Parkinson's, and ALS. These diseases do their damage at the molecular level, which only quantum computers can unravel and help fight. In the next chapter, we will investigate how quantum computers may reveal new insights about our immune system and eventually strengthen it.

CHAPTER 11

GENE EDITING AND
CURING CANCER

In 1971, with great fanfare, President Richard Nixon announced the War on Cancer. Modern medicine, he declared, would finally end this great scourge.

But years later, when historians evaluated this effort, the verdict was clear: cancer had won. Yes, there were incremental inroads in fighting it with surgery, chemotherapy, and radiation, but the number of cancer deaths remained stubbornly high. Cancer is still the second-leading killer in the U.S., next to cardiovascular diseases. Worldwide, it killed 9.5 million people in 2018.

The fundamental problem with the War on Cancer was that scientists did not know what cancer really was. There was a raging debate about whether this dreaded disease was caused by a single factor, or a confusing collection of them, such as diet, pollution, genetics, viruses, radiation, smoking, or just bad luck.

Several decades later, advances in genetics and biotechnology have finally revealed the answer. At the most fundamental level, cancer is a disease of our genes, but it can be triggered by environmental poisons, radiation, and other factors—or just plain bad luck. In fact, cancer is not one disease at all, but thousands of different types of mutations in our genes. There are now encyclopedias of the various types of cancers that cause healthy cells to suddenly proliferate and kill the host.

Cancer is an incredibly diverse and pervasive disease. It is found in mummies that are thousands of years old. The oldest medical reference to it dates back to 3000 BCE in Egypt. But cancer is found not just in humans. It is found throughout the animal kingdom. Cancer, in some sense, is the price we pay for having complex life-forms on earth.

To create a complex life-form, involving trillions of cells performing complicated chemical reactions in sequence, some cells have to die as new ones come to take their place, which allows the body to grow and develop. Many of the cells of a baby must eventually die to pave the way for the cells of an adult. This means that cells are genetically programmed to die by necessity, sacrificing themselves to create new complex tissues and organs. This is called apoptosis.

Although this programmed cell death is part of the body's healthy development, errors can sometimes turn off these genes accidentally, so the cell continues to reproduce and proliferate wildly. These cells cannot stop reproducing, and in that sense, cancer cells are immortal. In fact, that is why they can kill us, by growing uncontrollably and creating tumors that eventually shut down vital bodily functions.

In other words, cancer cells are ordinary cells that have forgotten how to die.

It often takes many years or decades for cancer to form. For example, if you had a severe sunburn as a child, you may get skin cancer at that very spot decades later. This is because it takes more than one mutation to cause cancer. It often takes years or decades for several mutations to accumulate, which will then finally disable the cell's ability to control reproduction.

But if cancer is so deadly, then why didn't evolution get rid of these defective genes millions of years ago by natural selection? The answer is that cancer mainly spreads after our reproductive years are over, so there is less evolutionary pressure to eliminate cancer genes.

We sometimes forget that evolution progresses through natural selection and chance. Therefore, while the molecular mechanisms that make life possible are indeed marvelous, they are the by-product of random mutations over billions of years of trial and error. Hence, we cannot expect our body to mount a perfect defense against deadly diseases. Given the bewildering number of mutations involved in cancer, it may take quantum computers to sift through this mountain of information and identify the root causes of the disease. Quantum computers are ideally suited to attack a disease that manifests in so many confusing ways. They may eventually give us an entirely new battleground in which to confront incurable diseases like cancer, Alzheimer's, Parkinson's, ALS, and others.

Liquid Biopsies

How do we know if we have cancer? Sadly, many times we don't. The signs of cancer are sometimes ambiguous or hard to detect. By the time a tumor forms, for example, there may be billions of cancer cells growing in the body. If a malignant tumor is found, almost immediately your doctor may recommend surgery, radiation, or chemotherapy. Sometimes, however, it's too late.

But what if you could stop the spread of cancer by detecting anomalous cells before a tumor forms? Quantum computers may play a key role in these endeavors.

Today, during a routine visit to the doctor's office, we take a blood test and might be given a clean bill of health. Yet later, the telltale signs of cancer may emerge. So you might ask yourself, why can't a simple blood test detect cancer?

This is because our immune system usually cannot detect cancer cells. They fly under the radar. Cancer cells are not foreign invaders easily recognized by our immune system. They are our own cells gone bad, and hence can elude discovery. Therefore, blood tests that analyze immune response can't recognize the presence of cancer.

But it has been known for more than 100 years that cancer tumors shed cells and molecules into bodily fluids. For example, cancer cells and molecules can be detected in blood, urine, cerebrospinal fluid, and even saliva.

Unfortunately, this is only possible if there are already billions of cancer cells growing in your body. By that point, surgery is usually required to remove the tumor. But recently, genetic engineering has finally given us the ability to detect cancer cells floating in our bloodstream or other bodily fluids. One day, this method may become sensitive enough to detect just a few hundred cancer cells, giving us years to act before a tumor forms.

But only within the last few years has it been possible for the average person to create an early warning system for cancers. One promising avenue of research is called a liquid biopsy, which is a fast, convenient, and versatile way of detecting cancer that may create a revolution in cancer detection.

"In recent years, the clinical development of liquid biopsies for cancer, a revolutionary screening tool, has created great optimism," write Liz Kwo and Jenna Aronson in the *American Journal of Managed Care.*

At present, liquid biopsies can detect up to fifty different types of cancer. A standard visit to the doctor may eventually be able to detect cancers years before they become lethal.

In the future, even the toilet in your bathroom may be sensitive enough to detect the signs of cancer cells, enzymes, and genes circulating in your bodily fluids, so that cancer becomes no more lethal than the common cold. Every time you go to the bathroom, you might be unwittingly tested for cancer. The "smart toilet" might be our first line of defense.

Although thousands of different mutations can cause cancer, quantum computers can learn to identify them so that a simple blood test may be able to detect scores of possible cancers. Perhaps our genome may be read on a daily or weekly basis and scanned by distant quantum computers to find any evidence of harmful mutations. This is not a cure for cancer, but it allows one to prevent it from spreading so it becomes no more dangerous than the common cold.

Many people ask the simple question, "Why can't we cure the common cold?" Actually, we can. But since there are over 300 rhinoviruses that can cause colds, and since they constantly mutate, it makes no sense to develop 300 vaccines to hit this moving target. We simply live with it.

This may be the future of cancer research. Instead of being a death sentence, it may eventually be viewed as a nuisance. Since there are so many cancer genes, it might be impractical to devise cures for all of them. But if we can detect them with quantum computers years before they spread, when they are just a small colony of a few hundred cancer cells, then it might be possible to stop their progression.

In other words, in the future, we may always have cancer, but perhaps only rarely will it kill anyone.

Sniffing Cancers

Another way to spot cancer in its early stages might be to use sensors to detect the faint odors given off by cancer cells. One day, perhaps your cell phone, with attachments that are sensitive to odors and connected to a quantum computer in the cloud, may help defend not just against cancer, but a variety of other diseases. Quantum computers would analyze the results of millions of "robotic noses" across the country to stop cancer in its tracks.

Analyzing odor is a proven diagnostic technique. For example, dogs are being used to detect the coronavirus at airports. While a typical PCR test for the virus may take a few days, specially trained dogs can give a 95 percent accurate identification within about ten seconds. This is already being used to screen passengers in the Helsinki airport and elsewhere.

Dogs have been trained to identify lung, breast, ovarian, bladder, and prostate cancer. In fact, dogs have a 99 percent success rate in detecting prostate cancer by sniffing a patient's urine sample. In one study, dogs could detect breast cancer with 88 percent accuracy and lung cancer with 99 percent accuracy.

The reason is that they have 220 million nasal scent receptors, while humans have only 5 million. So their sense of smell is many times more accurate than that of humans. It is so accurate that they can detect concentrations of one part per trillion, which is equivalent to detecting a single drop of liquid in twenty Olympic-sized swimming pools. And the area of their brain devoted to analyzing smells is much larger than its counterpart in human brains.

However, one drawback is that it takes a few months to train a dog to recognize the coronavirus or cancer, and there is a limited supply of such specially trained dogs. Could we perform these analyses with our own technology at a scale that might save millions of lives?

Soon after 9/11, I was invited by a TV company for a special luncheon to discuss the technologies of the future. I had the privilege of sitting beside an official from DARPA (Defense Advanced Research Projects Agency), a branch of the Pentagon famous for inventing the technology of the future. DARPA has a long history of spectacular success stories, such as NASA, the internet, the driverless car, and the stealth bomber.

So I asked him a question that had always bothered me: Why can't we develop sensors that detect explosives? Dogs can easily perform feats that our finest machines cannot.

He paused for a moment, and then he slowly explained to me the difference between dogs and our most advanced sensors. DARPA, in fact, had indeed studied this question carefully, and noted that the olfactory nerves of dogs are so sensitive that they can even pick up individual molecules of certain odors. Artificial sensors developed in our best laboratories cannot approach that sensitivity.

A few years after that conversation, DARPA sponsored a contest to see if laboratories could create a robotic nose like that of a dog.

One person who heard about that challenge was Andreas Mershin of MIT. He was fascinated by the near-miraculous ability of dogs to detect a variety of diseases and ailments. Mershin first got interested in this question when he was studying bladder cancer detection. One dog had persistently identified a particular patient as having cancer, even though he had been tested numerous times and was deemed cancer free. Something was wrong. The dog never changed its position. Finally, the patient agreed to be tested again, and he was found to have bladder cancer at a very early stage before it could be detected using standard laboratory tests.

Mershin wanted to replicate this astonishing success. His goal was to create a "nano-nose," which has microsensors capable of detecting cancers and other ailments, and then alerting you via your cell phone. Today, scientists at MIT and Johns Hopkins University have developed microsensors that are 200 times more sensitive than a dog's nose.

But because the technology is still experimental, it costs about $1,000 to analyze one sample of urine for cancer. Still, Mershin envisions the day when this technology will be as common as the camera in your cell phone. Because of the sheer mass of data that could come pouring in from hundreds of millions of cell phones and sensors, only quantum computers would have the ability to process this treasure trove of data. It then could use artificial intelligence to analyze the signals, locate any cancerous markers, and send the information back to you, perhaps years before a tumor forms.

In the future, there may be several ways to effortlessly and silently detect cancer before it poses a serious threat. Liquid biopsies and odor detectors may be able to send data to a quantum computer, which could identify scores of different types of cancer. In fact, the word "tumor" may disappear from common discourse in the English language, in the same way we no longer talk about "bloodletting" or "leeches."

But what happens if cancer has already formed? Can quantum computers help cure cancer once it has begun to attack the body?

Immunotherapy

At present, there are at least three main ways in which to attack cancer once it is detected: surgery (to cut out the tumor), radiation (to kill cancer cells with X-rays or particle beams), and chemotherapy (to poison cancer cells). But with the emergence of genetic engineering, a new form of therapy is gaining widespread use: immunotherapy. There are several versions of this treatment, but, in general, they all try to enlist the help of the body's own immune system.

Cancer cells, as discussed, unfortunately cannot easily be identified by the body's immune system. The body's T and B cells, for example, are programmed to identify and later kill a vast number of foreign antigens, while cancer cells are not part of the library of antigens that the white blood cells can recognize. Thus, they fly under the radar of our immune system. The trick is to artificially boost the power of our own immune system to recognize and attack the cancer.

In one method, the genome of the cancer is sequenced, so doctors know precisely the type of cancer being studied and how it is developing. Next, white blood cells are extracted from our blood while genes from the cancer are processed. The genetic information of the cancer is then inserted into the white blood cells via a virus (which has been previously rendered harmless). Now the white blood cells have been reprogrammed to identify these cancer cells. Finally, the recalibrated white blood cells are injected back into the body.

So far, this method shows great promise in terms of attacking incurable forms of cancer, even in late stages when it has spread across the body. Some patients who were told that their case was hopeless have suddenly and dramatically seen their cancers disappear.

Immunotherapy has been used for cancer of the bladder, brain, breast, cervix, colon, rectum, esophagus, kidney, liver, lung, lymph, skin, ovary, pancreas, prostate gland, bone, stomach, and for leukemia, all with varying degrees of success.

But there are drawbacks. This method is only available for some cancers, and there are thousands of different types. But also, because the genetics of the white blood cells have been altered artificially, sometimes the modification is not perfect. This may cause unwanted side effects. These side effects may, in fact, sometimes be fatal.

Quantum computers, though, might be able to help perfect this therapy. Eventually, quantum computers may be able to analyze this mass of raw data to identify the genetics of each cancer cell. Such a monumental task would overwhelm a classical computer. Each person in the country would have their genome read, silently and efficiently, several times a month through an analysis of their bodily fluids. Their entire genome would be sequenced, cataloguing over 20,000 genes per person. Then this would be compared to the thousands of possible cancer genes that have been studied. A vast infrastructure of quantum computers would be required to analyze this mass of raw data. But the benefits would be enormous: the diminution of this dreaded killer.

Paradox of the Immune System

There has been a long-standing mystery about the immune system. In order for the body to destroy invading antigens, it must first be able to identify them. Since there are essentially an unlimited number of possible viruses and bacteria, how does the immune system tell the difference between the dangerous ones and the friendly ones? How does it know the difference when it has never encountered a particular disease before? This is like the police knowing whom to arrest in a crowd of people they've never seen before.

At first, it seems impossible. There are in principle an infinite number of different types of diseases, so it isn't clear how the immune system could magically find just the right ones.

But evolution has devised a clever way to solve this problem. The B white blood cell, for example, contains Y-shaped antigen receptors that protrude from its cell wall. The goal of the white blood cell is to latch the tips of its Y receptor to a dangerous antigen so that it can either be destroyed or marked for later destruction. This is how it identifies threatening antigens.

When the white blood cell is born, the genetic codes in the tips of the Y receptors that match the receptors to specific antigens are randomly mixed. This is the key. So in principle almost *all* the codes that the body may ever encounter are already contained within the various random Y receptors, both good and bad. (To appreciate how a small number of amino acids can create huge numbers of genetic codes, consider a hypothetical example. We start with the fact that there are 20 different amino acids in the human body. Let's say we create a chain of 10 amino acids, with 20 possible amino acids in each slot. Then there are $20 \times 20 \times 20 \times \ldots = 20^{10}$ possible random arrangements of amino acids. Compare this to the actual number of possible B cell receptors, which has about 10^{12} different possible combinations. This astronomical number contains almost all the possible antigens that it may ever encounter.)

Once the Y receptors are all randomized, however, the receptors that contain the genetic codes of the body's own amino acids are gradually removed. Left behind are Y receptors, therefore, that only contain the genetic code of dangerous antigens. In this way, the Y receptors can attack dangerous antigens even if they have never encountered them before.

So this is like the police trying to find a criminal within a huge crowd of people. First the police eliminate all the people who were previously known to be innocent. Then the police know that the criminals might be among those left behind.

Because we live in an invisible ocean of billions of bacteria and viruses, the system works surprisingly well. However, sometimes it backfires. For example, sometimes when deleting the genetic codes that are found in the body, the body does not eliminate them all. Some of the good codes are then left behind, to be attacked by the immune system. In other words, if the police don't eliminate all the innocent suspects, some innocents are accidentally left behind. Then when it is time to interrogate the suspects, some innocents are suspected as well.

This means that the body will then attack itself, creating a host of autoimmune diseases. Perhaps this is why we have rheumatoid arthritis, lupus, type 1 diabetes, multiple sclerosis, etc.

Sometimes, the reverse happens. The immune system not only removes the good codes, it eliminates some bad codes as well by accident. Then the immune system cannot identify the dangerous ones, which can cause disease.

This is what might happen sometimes with some types of cancer, when the body is unable to detect the antigens with the wrong genes.

The entire process of identifying dangerous antigens is purely a quantum mechanical one. Digital computers are incapable of reproducing the complex sequence of events that must be played out at the molecular level in order for the immune system to work properly. But quantum computers may be powerful enough to unravel, molecule for molecule, how the immune system does its magic.

CRISPR

The therapeutic applications of quantum computers may be increased when combined with a new technology called CRISPR (clustered regularly interspaced short palindromic repeats), which allows scientists to cut and paste genes. Quantum computers can be used to identify and isolate complex genetic diseases, and CRISPR might be used to cure them.

Back in the 1980s, there was enormous enthusiasm about gene

therapy, i.e., repairing broken genes. There are at least 10,000 known genetic diseases afflicting the human race. There was a belief that science would enable us to rewrite the code of life, correcting the mistakes of Mother Nature. There was even talk that gene therapy might be able to enhance the human race as well, improving our health and intelligence at the genetic level.

Much of the early research was focused on an easy target: attacking genetic diseases that are caused by a misspelling of a few letters in our genome. For example, sickle cell anemia (which afflicts many African Americans), cystic fibrosis (which affects many northern Europeans), and Tay-Sachs (which affects Jewish people) are caused by the misspelling of one or a few letters in our genome. There was hope that doctors would be able to cure these diseases by simply rewriting our genetic code.

(Because of intermarrying, these genetic diseases were so prevalent in the royal families of Europe that historians have written that they even affected world history. King George III of England suffered from a genetic disease that rendered him mad. Historians have speculated that his insanity may have led to the American Revolution. Also, the son of Nicholas II of Russia was afflicted with hemophilia, which the royal family believed could only be treated by the mystic Rasputin. This paralyzed the monarchy and delayed needed reforms, which might have contributed to the Russian Revolution of 1917.)

These genetic engineering trials were conducted in a similar way to immunotherapy. First the desired gene was inserted into a harmless virus, modified so that it could not attack its host. Then the virus would be injected into the patient, so the patient was infected with the desired gene.

Unfortunately, complications soon arose. For example, the body would often recognize the virus and attack it, causing unwanted side effects for the patient. Many of these hopes for gene therapy were dashed in 1999 when a patient died after a trial. Funds began to dry up. Research programs were drastically scaled back. Trials were reexamined or halted.

But more recently, researchers had a breakthrough when they began to look closely at how Mother Nature attacks viruses. We sometimes forget that viruses attack not only people, but also bacteria. So doctors asked a simple question: How do bacteria defend themselves against the onslaught of viruses? Much to their surprise, they found that, over millions of years, bacteria have devised ways to cut up the genes of the invading virus. If a virus tries to attack a bacteria, the bacteria may counterattack by releasing a barrage of chemicals that split the genes of the virus at precise points, thereby stopping the infection. This powerful mechanism was isolated and then used to sever viral genetic codes at desired points. The Nobel Prize was given to Emmanuelle Charpentier and Jennifer Doudna in 2020 for their pioneering work in perfecting this revolutionary technology.

This process has been compared to word processing. In the old days, typewriters had to type each letter successively, which was a painful, error-ridden procedure. But with word processors, it was possible to write a program that could allow one to edit entire manuscripts by removing and rearranging its pieces. Similarly, one day, perhaps CRISPR technology can be applied to genetic engineering, which has had mixed success over the years. This would open the floodgates to genetic engineering.

One particular target for gene therapy might be the gene p53. When mutated, it is involved in about half of all common cancers, such as cancer of the breast, colon, liver, lung, and ovaries. Perhaps one reason why it is so vulnerable to becoming cancerous is that it is an exceptionally long gene, and hence there are many sites on it where mutations may develop. It is a tumor-suppressor gene, which makes it vital in stopping the growth of cancers. For that reason, it is often called "The Guardian of the Genome."

But when mutated, it becomes one of the most common underlying genes in human cancers. In fact, mutations at specific sites are often correlated with specific cancers. For example, longtime smokers often develop cancers at three specific mutations along p53, which may be used to prove that this person's lung cancer most likely came from cigarette smoke.

In the future, using advances in gene therapy and CRISPR, one might be able to fix the misspellings in the p53 gene using immunotherapy and quantum computers, and thus cure many forms of cancer.

We recall that immunotherapy has side effects, including in rare cases the death of the patient. Part of the reason for this is that the cutting and pasting of cancer genes is imprecise. P53, for example, is a very long gene, so errors in cutting this gene may be common. Quantum computers may help reduce these lethal side effects. They could potentially decipher and map the molecules within the genes of a certain cancer cell. Then CRISPR may be able to accurately cut the gene at precise points. So using a combination of gene therapy, quantum computers, and CRISPR may make it possible to cut and splice genes with ultimate precision, reducing the problem of lethal side effects.

CRISPR Gene Therapy

Clara Rodríguez Fernández writes in *Labiotech:* "In theory, CRISPR could let us edit any genetic mutation at will to cure any disease with a genetic origin." Genetic diseases that involve a single mutation are the ones being targeted first. She adds, "With over 10,000 diseases caused by mutations in a single human gene, CRISPR offers hope to cure all of them by repairing any genetic error behind them." In the future, as the technology develops, genetic diseases caused by multiple mutations in several genes may be studied.

For example, here is a list of some of the genetic diseases currently being treated by CRISPR:

1. Cancer

At the University of Pennsylvania, scientists were able to use CRISPR to remove three genes that allow cancer cells to evade the body's immune system. Then they added another gene that can help the immune system recognize tumors. The scientists found that the method was safe, even when used on patients with advanced cancer.

In addition, CRISPR Therapeutics is running a test on 130 patients suffering from blood cancer. These patients are being treated with immunotherapy, which uses CRISPR to modify their DNA.

2. Sickle Cell Anemia

CRISPR Therapeutics is also harvesting bone marrow stem cells from patients suffering from sickle cell anemia. CRISPR is then altering these cells to produce fetal hemoglobin. These treated cells are then inserted back into the body.

3. AIDS

A small number of individuals are born with a natural immunity to AIDS because of a mutation in their CCR5 gene. Normally, the protein made by this gene creates an entry point for the AIDS virus to enter a cell. However, in these rare individuals, the CCR5 gene is mutated so the AIDS virus cannot penetrate into a cell. For people without this mutation, scientists are deliberately editing the CCR5 gene with CRISPR so that the virus cannot enter their cells.

4. Cystic fibrosis

Cystic fibrosis is a relatively common respiratory disease; individuals suffering from it rarely live beyond forty years of age. It is caused by a mutation in the gene CFTR. In the Netherlands, doctors were able to use CRISPR to repair that gene without causing side effects. Other groups, such as Editas Medicine, CRISPR Therapeutics, and Beam Therapeutics are also planning to treat cystic fibrosis with CRISPR.

5. Huntington's disease

This genetic disease often causes dementia, mental illness, impaired cognition, and other debilitating symptoms. It is believed that some of the women persecuted at the Salem witch trials in 1692 suffered from this disease. It is the result of a repetition of the Huntington gene along the DNA. Scientists at the Children's Hospital of Philadelphia are using CRISPR to treat this disease.

While diseases caused by minimal mutations make relatively easy targets for CRISPR, diseases like schizophrenia may involve a large number of mutations, plus interactions with the environment. This is another reason why quantum computers may be required.

To understand how these mutations can create an illness at the molecular level may necessitate the full power of quantum computers. Once we know the molecular mechanism by which certain proteins cause genetic diseases, then we can modify them or find more effective treatments.

Peto's Paradox

But this also raises a paradox about cancer. Biologist Richard Peto of Oxford noticed something odd about elephants. Because of their massive size, one would expect that they would have more cancers than much smaller animals. After all, a larger mass means more cells are constantly dividing and introducing the possibility for genetic errors, like cancer. But surprisingly enough, elephants have a relatively low cancer rate. This became known as Peto's paradox.

When analyzing the animal kingdom, we see this everywhere. The rate of cancer often does not correspond to body weight. Later, it was found that elephants have twenty copies of the p53 gene, while we humans only have one copy. It is believed that these extra copies of p53 work with another gene called LIF to give elephants the advantage against cancer. So genes like p53 and LIF are thought to work to suppress the cancers in large animals.

But this might not be the whole story. For example, whales have only one copy of p53 and one version of LIF, yet they have a low rate of cancer. This means that whales probably have other genes, which have not yet been found by scientists, that protect them against cancer. In fact, it is believed that there could be numerous genes that prevent large animals from falling victim to high rates of cancer. Certain sharks may also have some genetic advantage conferred on them by evolution. Greenland sharks can live for up to 500 years, which is probably made possible by a still unknown gene.

"The hope is that by seeing how evolution has found a way to prevent cancer, we could translate that into better cancer prevention. Every organism that evolved large body size has a different solution to Peto's paradox. There's a bunch of discoveries that are just waiting for us out there in nature, where nature is showing us the way to prevent cancers," says Carlo Maley, who has studied the p53 gene in the animal kingdom. And quantum computers may prove instrumental in finding these mysterious anti-cancer genes.

There are many ways in which quantum computers may help in the war against cancer. One day, liquid biopsies may be able to detect cancer cells years to decades before tumors form. In fact, one day quantum computers could make possible a gigantic national repository of up-to-the-minute genomic data, using our bathrooms to scan the entire population for the earliest signs of cancer cells.

But if cancer does form, quantum computers might enable modifications to our immune system that would allow it to attack hundreds of different types of cancer. A combination of gene therapy, immunotherapy, quantum computers, and CRISPR could potentially cut and paste cancer genes with great molecular precision, helping reduce the often lethal side effects of immunotherapy. Further, perhaps a handful of genes, like p53, are involved in the vast majority of these cancers, so gene therapy combined with new insights from quantum computers may be able to stop them in their tracks.

All these breakthroughs in treating cancer, such as liquid biopsies and immunotherapy, prompted President Joseph Biden in 2022 to declare a Cancer Moonshot, a national goal to reduce the cancer death rate by at least 50 percent over the next twenty-five years. Given the rapid advances in biotechnology, this is certainly an achievable goal.

Although we may have the ability to completely cure an increasing number of cancers using this technology, we probably will still suffer from some forms of cancer simply because there are so many ways in which a cancer can form. But in the future, we may treat cancer like the common cold, as a preventable nuisance. But another powerful combination of new technologies, which we explore in the next chapter, might give us a line of defense against disease. AI and quantum computers may also endow us with the ability to create designer proteins, out of which our bodies are made. Together, they may enable us to cure incurable diseases and reshape life itself.

CHAPTER 12

AI AND QUANTUM COMPUTERS

C an machines think?

This was the question that dominated the historic 1956 Dartmouth Conference, which gave birth to an entirely new field of science, dubbed "artificial intelligence." It started with a bold proposal that read: "An attempt will be made to find how to make machines use language, form abstractions and concepts, solve kinds of problems now reserved for humans and improve themselves." They predicted that "significant advance can be made . . . if a carefully selected group of scientists work on it together for a summer."

Many summers later, some of the world's brightest scientists are still doggedly working on this problem.

One of the leaders of that conference was MIT professor Marvin Minsky, who has been called the Father of Artificial Intelligence.

When I asked him about that period, he said those were heady times. It seemed that within just a few years it would be possible to match the intelligence of a human with a machine. Perhaps it was only a matter of time before robots would pass the Turing test.

It seemed that every year, new breakthroughs were happening in the field of AI. For the first time, digital computers could play checkers and even beat humans in simple games. There were computers that could solve algebra problems like a schoolchild. Mechanical arms were designed that could identify and then pick up blocks. At the Stanford Research Institute, scientists built Shakey, a boxlike minicomputer placed on treads, with a camera on top. It could be programmed to roam around a room and identify the objects in its path. It could navigate by itself and avoid obstacles. (Its name came from all the noise it made while lumbering across the floor.)

The media went crazy. The mechanical man was being born right before our eyes, they cried. Headlines in science magazines heralded the arrival of the household robot, which could vacuum the floor, do the dishes, and relieve us of housework. Robots would one day become babysitters or even trusted members of the family. Even the military was opening its checkbook and funding robots for battle-field use, like the Smart Truck, which could one day travel on its own, perform reconnaissance behind enemy lines, rescue injured soldiers, and then report back to base, all by itself.

Historians began to write that we were on the verge of fulfilling an ancient dream. The Greek god Hephaestus created a fleet of robots to do chores around his castle. Pandora, who opened a magic box and unknowingly unleashed disaster upon the human race, was actually a robot built by Hephaestus. And even the polymath Leonardo da Vinci in 1495 built a mechanical knight that could maneuver its arms, stand, sit, and raise its visor, operated by a series of hidden cables and pulleys.

But then "AI winter" set in. In spite of all the breathless press releases, AI had been oversold to the media, and dark clouds of pessimism set in. Scientists began to realize that their AI devices were one-trick ponies. They could only do one simple task each. Robots were still clumsy devices that could barely navigate around a room. The idea of creating an all-purpose machine to match the intelligence of a human seemed impossibly advanced.

The military began to lose interest. Funding dried up and investors lost their shirts. Since that time, there have been several AI winters, where the boom-bust cycle would generate enormous enthusiasm and shameless publicity, only to come crashing down. Scientists had to face the harsh reality that AI was harder to develop than they thought.

Given the fact that Marvin Minsky had seen so many AI winters come and go, I asked him if he had any prediction of when a robot may match or surpass human intelligence. He smiled and told me that he no longer made predictions like that about the future. He was no longer in the business of peering into crystal balls. Too many times, he admitted, people let their enthusiasm run away with them.

The problem, he told me, is that AI researchers suffer from what he called "physics envy," the desire to find a single, unifying, over-arching theme to AI. Physicists, he said, search for a single unified field theory that will give a coherent, elegant picture of the universe, but AI is different. It's a messy patchwork with too many divergent and even conflicting pathways given to us by evolution.

New ideas and new strategies have to be explored. One promis-ing avenue might be to marry AI and quantum computers, to merge the power of these two disciplines to tackle the problem of artificial intelligence. In the past, AI was wedded to digital computers, so there were frustrating limits to what a computer could do. But AI and quantum computers complement each other. AI has the ability to learn new, complex tasks, and quantum computers can provide the computational muscle it needs.

A quantum computer may have formidable power, but it does not necessarily learn from its mistakes. But a quantum computer equipped with neural networks will be able to improve its calcula-tions with each iteration, so it can solve problems faster and more efficiently by finding new solutions. Similarly, AI systems may be equipped to learn from their mistakes, but their total calculational capacity may be too small to solve very complex problems. So an AI backed up with quantum computer calculational power could tackle more difficult problems.

In the end, the union of AI and quantum computers may open up entirely new avenues for research. Perhaps the key to artificial intelligence lies in the quantum theory. In fact, the merger of the two may revolutionize every branch of science, alter our lifestyle, and radically change the economy. AI will give us the ability to create learning machines that can begin to mimic human abilities, while quantum computers may provide the calculational power to finally create an intelligent machine.

As the CEO of Google, Sundar Pichai, has said, "I think AI can accelerate quantum computing, and quantum computing can accelerate AI."

Learning Machines

One scientist who has thought long and hard about the future of AI is Rodney Brooks, former director of MIT's Artificial Intelligence Laboratory, which was founded by Marvin Minsky.

Brooks believes AI may have been conceived too narrowly. For example, he told me, consider a fly. It can perform miraculous feats of navigation that outperform our finest machines. All by itself, it can deftly fly around a room, maneuver, avoid obstacles, locate food, find mates, and hide, all with a brain no bigger than a pinpoint. It is truly a marvel of biological engineering.

How can that be? How can Mother Nature create a flying machine that would put our finest aircraft to shame?

He began to realize that maybe the field of AI was asking the wrong questions back in 1956. Back then, it was assumed that the brain was a Turing machine of some sort, a digital computer. You write down the complete rules for chess, walking, algebra, etc. into one gigantic piece of software, and then you insert it into the digital computer, and suddenly it begins to think. "Thinking" was reduced to software and hence the basic strategy was clear: write increasingly sophisticated software to guide the machine.

A Turing machine, we recall, has a processor that carries out the commands fed into it. It is only as intelligent as the programming it implements. So a walking robot has to have all of Newton's laws of motion programmed into it in order to guide the motion of its limbs, microsecond by microsecond. This requires gigantic computer programs, with millions of lines of computer code, to simply walk across the room.

AI machines up to then, Brooks told me, were based on programming all the laws of logic and motion from the very start, which turned out to be an arduous task. This was called the top-down approach, when robots were programmed to master everything from the very beginning. But the robots designed this way were pathetic. If you take Shakey or an advanced military robot of the period and put it in the forest, what does it do? Most likely, it gets lost or falls over. Yet the smallest insect with its minuscule brain can whiz around the area, find food, mates, and shelter while our robot flails helplessly on its back.

This is not how Mother Nature has designed her creatures.

In nature, Brooks realized, animals are not programmed to walk from the start. They learn the hard way, by putting one leg in front of the other, falling down, and doing it again. Trial and error is the way of nature.

This goes back to the advice every music teacher gives to their promising student. How do you get to Carnegie Hall? Answer: practice, practice, practice.

In other words, Mother Nature designs creatures that are pattern-seeking learning machines, using trial and error to navigate the world. They make mistakes, but with each iteration, they come closer to success.

This is a bottom-up approach, and it starts with nothing but bumping into things. For example, babies learn by mimicking adults. If you put a tape recorder in a crib at night, you will hear babies babble constantly. What they are actually doing is endlessly practicing making the sounds they hear over and over again, until they can duplicate them correctly.

So, guided by this insight, Brooks created a fleet of "insectoids" or "bugbots." They learn how to walk as Mother Nature intended, by bumping into things. Soon, tiny, insect-like robots were crawling over the floor at MIT, bumping into things, but outsmarting the more clumsy, traditional robots that follow strict rules but gouge the wallpaper as they lumber by. Why reinvent the wheel?

Brooks told me, "When I was a kid, I had a book that described the brain as a telephone switching network. Earlier books described it as hydrodynamic system or a steam engine. Then in the 1960s, it became a digital computer. In the 1980s, it became a massively parallel digital computer. Probably there's a kid's book out there somewhere that says the brain is just like the World Wide Web."

So maybe the brain is actually a pattern-seeking learning machine, based on what are called neural networks. In computer science, neural networks exploit something called Hebb's rule. One version of this rule states that by constantly repeating a task and learning from previous mistakes, each iteration gets closer to the correct path. In other words, the correct electrical pathways for that task get reinforced in the brain of the AI system after repeated iterations.

For example, when a learning machine tries to identify a cat, it is not given the mathematical description of the basic features of a cat. Instead, it is shown scores of pictures of cats, in all sorts of situations—sleeping, crawling, hunting, jumping, etc. Then the computer figures out by itself what a cat looks like in different environments by trial and error. This is called deep learning.

The successes of the deep learning approach are quite remarkable. Google's AlphaGo, an AI designed to play the ancient board game Go, was able to beat the world champion in 2017. This was a remarkable feat, since there are 10^{170} possible positions in Go on a 19 x 19 board. This is more than all the atoms in the known universe.

AlphaGo learned how to play not just by facing the top human players, but also by playing against itself, where it could run through games at nearly the speed of light.

Commonsense Problem

Learning machines or neural networks may eventually solve one of the most stubborn problems in artificial intelligence: the "commonsense problem." Things that humans take for granted, that even a child can understand, are beyond the capability of our most advanced computers. Until a robot can solve the commonsense problem, they will be unable to function in human society.

For example, a digital computer may not understand a simple set of observations, such as:

- Water is wet, not dry
- Mothers are older than their daughters
- Strings can pull, but cannot push
- Sticks can push, but cannot pull

In an afternoon, it is easy to write down scores of "obvious" facts about our world that are beyond the understanding of digital computers. This is because computers do not experience the world as we do.

Children learn these commonsense facts because they bump into these things. They learn by doing. They know that mothers are older than their daughters because they have seen them through their experience. But a robot is a clean slate, with no prior understanding of their environment.

As we discussed with the top-down approach, scientists have tried to program common sense into the software of a computer. Instantly, it would know how to navigate and function in human society. However, all such attempts have ended in failure. There are simply too many commonsense notions, which even a four-year-old child understands, that are beyond the reach of our digital computers.

Thus, perhaps the merger of the top-down and bottom-up approaches, and the merger of AI with quantum computers, will fulfill the dream of the first AI researchers and pave the road to the future.

As Moore's law slows down, as we have seen, due to transistors approaching the size of atoms, microchips will inevitably be replaced by more advanced computers, such as quantum computers.

AI, for its part, has stalled because of the lack of computer power. Its capabilities in machine learning, pattern recognition, search engines, and robotics are all constrained by this limit. Quantum computers can vastly accelerate progress in each of these areas because they can process vast amounts of information simultaneously. While digital computers compute one bit at a time, quantum computers compute on a huge array of qubits at the same time, thereby magnifying their power exponentially.

So we see how AI and quantum computers can cross-fertilize each other. Quantum computers can benefit from being able to learn new tasks, as in a neural network, and AI can benefit from the vast computational firepower of quantum computers.

Protein Folding

AI deep learning systems are now tackling one of the greatest problems in all of biology and medicine: to decode the secret of protein molecules. Although DNA contains the instructions for life, it is the proteins that actually do the yeoman's work of making the body function. If we compare our body to a construction site, the DNA contains the blueprints, but the proteins do the heavy lifting of the foremen and construction workers. A blueprint is useless without an army of workers to carry it out.

Proteins are the workhorses of biology. Not only do they make up the muscles that energize our body, they also digest our food, attack germs, regulate our bodily functions, and do many other crucial tasks. So biologists have wondered: How does the protein molecule perform all these miraculous functions?

In the 1950s and 1960s, scientists used X-ray crystallography to map the shape of a number of protein molecules, which are made of exactly twenty amino acids arranged in long strings creating complex tangles. Much to their surprise, scientists found that it is the shape of the protein molecule that makes their magic possible. Scientists say that in this case "function follows form," i.e., it is the shape of a protein molecule, with all its intricate knots and twirls, that creates the characteristic properties of that protein.

For example, consider the Covid-19 virus, which we know is shaped like the corona of the sun, with many protein spikes radiating out from its surface. These spikes are like keys that open the specific "locks" located on the surface of our lung cells. By opening these locks, the spike protein can then inject their genetic material into our lung cells, where they promptly make numerous copies of themselves. Then the cell dies, releasing these deadly viruses to infect even more healthy lung cells. These spikes are the reason why the world economy nearly crashed in 2020–22.

So the shape of the protein, more than anything else, determines how that molecule behaves. If one knew the shape of each protein molecule, then we would be one giant step closer to understanding how it works.

This is the "protein folding problem," the task of mapping the shape of all important proteins, and it may unlock the secret of many incurable diseases.

X-ray crystallography has been the key to determining the shape of a protein molecule, but it is a long, tedious process. Scientists begin by first chemically isolating and purifying the proteins they want to analyze, which then have to be crystallized. The crystallized protein is inserted into an X-ray diffraction machine, which shoots

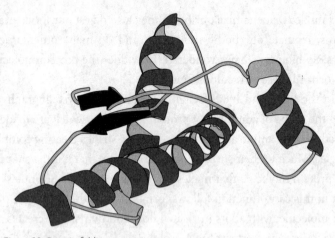

Figure 10: Protein folding
Proteins consist of a long string of twenty amino acids, which can fold up in complex ways. The shape of the folded protein molecule determines how it functions. Quantum computers may allow scientists to analyze and then create entirely new proteins with strange, but useful properties, creating a new branch of biology.

X-rays through the crystal and forms an interference pattern on photographic film. At first, the X-ray photograph looks like a hopeless jumble of dots and lines. But using intuition, luck, and physics, scientists try to decipher the structure of the protein from the X-ray pictures.

Birth of Computational Biology

So one of the goals of an emerging field called computational biology is to use computers to unravel the 3D structure of a protein just by looking at its chemical constituents. Perhaps all the years of hard work understanding the structure of a protein molecule could be done by pressing a button on a computer running an AI program.

To help spur research in this difficult but crucial area, scientists tried a new strategy. They created a contest called the CASP (Critical Assessment of Structure Prediction), to see who had the best computer program to crack the protein folding problem.

This was a turning point, because it gave young scientists an exciting, concrete goal. They could earn fame and the recognition of their peers by using AI to crack the protein folding problem, which could lead to therapies that would save thousands of lives.

The rules for the contest were simple. You were given the barest clues about the nature of a certain protein, such as the sequence of amino acids. Then it was up to your computer program to fill in all the details about how it folds up. One way to approach the problem was to use the principle of least action pioneered by Richard Feynman. You recall that Feynman, as a high school student, could determine which path a ball took by minimizing its action (its kinetic energy minus its potential energy).

You can apply the same method with protein molecules. The goal is to find the configuration of amino acids that creates the lowest energy state. This process has been likened to descending a mountain to find the lowest point in a valley. First, you take small, tentative steps in all directions. Next, you move only in the direction that slightly lowers your height. Then you start all over again and take the next step, seeing if you can lower yourself even further, until you reach the bottom of the valley.

In the same way, you can find the arrangement of amino acids that has the lowest energy. Here is one way to do this:

Before you begin, you make a series of approximations. Because a molecule has many wave functions describing electrons and nuclei all interacting with each other in complex ways, the calculation quickly exceeds the ability of a conventional computer. So you simply drop a number of complex terms that are relatively small (e.g., the interaction of electrons with heavy nuclei, and certain interactions between electrons) and hope that this does not create too many errors.

Now that you've set up the program, you first connect the various amino acids to each other in a long string. This creates a skeleton or a "toy model" of what the protein molecule might look like. Since you know the bonding angles when certain atoms connect with each other, this gives you a rough, initial approximation of what the protein might look like.

Second, you calculate the energy of this configuration of amino acids, because you know the energy of its various charges and how the bonds can move.

Third, you twist and turn these bonds, to see if the new configuration increases or decreases the energy of the protein. This is like the tentative steps you take on the mountain, feeling for the step that lowers your height.

Fourth, you discard all the configurations that increase the energy, keeping only the ones that lower it. The computer "learns" by trial and error how the movement of atoms can reduce the energy of the molecule.

And last, you start all over again by twisting the chemical bonds or rearranging the amino acids. With each iteration, you lower the energy by fiddling with the location and position of the amino acids until you finally reach the configuration of lowest energy.

Normally, this process of constantly adjusting the position of atoms would be impossible for a digital computer. But since you start with a series of approximations and drop complex terms that are relatively small, a computer can solve the simplified version within a matter of hours or days.

At first, the results were laughable. When comparing the shape of the molecule predicted by a computer, with the actual shape given by X-ray crystallography, the computer models were wildly off. But as the years went by, computer learning programs became more powerful, and the models became more precise.

By 2021, the results were spectacular. Even with all these approximations, the computer company DeepMind, which is affiliated with Google and developed AlphaGo, announced that their AI program, called AlphaFold, had deciphered the rough structure of an astounding number of proteins: 350,000. Furthermore, it identified 250,000 shapes that were previously unknown. It deciphered the 3D structure of all the 20,000 proteins listed in the Human Genome Project. It even unraveled the structure of proteins found in the mouse, fruit fly, and the *E. coli* bacterium. Later, the creators of DeepMind announced that they will soon release a database of more than 100 million proteins, which includes every protein known to science.

What is also remarkable is that, even with all these approximations, their final results roughly matched the results from X-ray crystallography. Despite dropping various terms in the Schrödinger wave equation, they were able to get surprisingly good results.

"We have been stuck on this one problem—how do proteins fold up—for nearly 50 years. To see DeepMind produce a solution for this, having worked personally on this problem for so long and after so many stops and starts, wondering if we'd ever get there, is a very special moment," says John Moult, cofounder of CASP.

This bonanza of information has already had immediate consequences. For example, it is being used to identify twenty-six different proteins found in the coronavirus, with the hope of finding its weak spots and creating new vaccines. In the future it should be possible to quickly find the structure of thousands of crucial proteins. "We were able to design coronavirus-neutralizing proteins in several months. But our goal is to do this kind of thing in a couple of weeks," says David Baker of the Institute for Protein Design at the University of Washington.

But this is just the beginning. As we have stressed, function follows form. That is, the way in which proteins do their job is determined by their structure. In the same way that a key fits into a keyhole, a protein performs its magic by somehow latching on to another molecule.

But uncovering how proteins fold was the easy part. Now begins the hard part, using quantum computers to determine the complete structure of a protein, without all the approximations, and how a particular protein fits together with other molecules so that it can perform its function, such as providing energy, acting as a catalyst, merging with other proteins, joining with other proteins to create new structures, splitting apart other molecules, and much else besides. So protein folding is just the first step in a long journey that contains the secrets of life itself.

In the future, the understanding of the protein folding program will progress in several stages, similar to the stages in the creation of genomics:

Stage One: Map the Folded Proteins

We are currently in Stage One, creating a huge dictionary, with hundreds of thousands of entries corresponding to the folding of various proteins. Each entry in this dictionary is a picture of the individual atoms that combine to make up a complex protein. These diagrams, in turn, came from studying X-ray photographs. This gigantic book has all the correct spelling of each protein, but is largely empty, without any definitions. It is based on a series of approximations that allow digital computers to make this calculation. It is rather surprising that, with so many approximations, scientists are still able to get such accurate results.

Stage Two: Determining the Proteins' Function

In the next stage, which we are now entering, scientists will try to determine how the geometric form of the protein molecule determines its function. AI and quantum computers will be able to identify how certain atomic structures in a folded protein can allow it to perform certain functions in the body. Eventually, we will have a complete description of bodily functions and how they are controlled by proteins.

Stage Three: Create New Proteins and Medicines

The final step is to use this dictionary of proteins to create new, improved versions, which will allow us to develop new medicines and therapies. To do this, we will have to abandon the approximations and solve for the actual quantum mechanics of molecules. Only quantum computers can achieve this.

Evolution has created a treasure trove of proteins by purely random interactions to carry out various tasks. However, it took billions of years to do this. Using the memory of a quantum computer as a "virtual laboratory," it should be possible to improve on evolution and design new proteins to improve their function in the body.

This process has a wide range of applications, including finding entirely new drugs. For starters, some people have envisioned how this would help clean up the environment. The simplest ongoing example is the work of scientists trying to find ways to break down the 150 million tons of soda bottles found in oceans, in waste dumps, and in your backyard. The key would be to use this protein database to examine the 3D shape of certain proteins, the enzymes capable of splitting apart molecules of plastic and rendering them harmless. This work is already being done at the Centre for Enzyme Innovation at the University of Portsmouth, England.

This may also have immediate medical applications, because a number of incurable diseases are linked to misfolded proteins. One promising avenue is to understand the nature of prions, which are potentially linked to a host of incurable diseases that can affect the elderly, such as Alzheimer's, Parkinson's, and ALS. So the clue to finding cures for these incurable diseases might come from quantum computers.

The frontiers of medicine, incurable diseases, may be the next battlefront for quantum computers.

Prions and Incurable Diseases

Traditionally every textbook says that diseases are spread by bacteria and viruses.

But that is probably not the whole story. It has been known for centuries that animals are subject to strange diseases unlike those affecting humans. Sheep with scrapie act strange, run their backs against posts, and refuse to eat. It is an incurable disease and always fatal. Mad cow disease (bovine spongiform encephalopathy) is a similar disease affecting cattle, in which they have trouble walking, become nervous, and even turn violent.

In humans, there is an exotic disease called kuru found among certain tribes in New Guinea. There, some tribes perform a funeral ceremony that involves eating the brains of the deceased. Some of these individuals suffered from dementia, mood swings, difficulty walking, and other symptoms because of a new disease that was found in the brains of their relatives.

Stanley B. Prusiner of the University of California at San Francisco went against the tide of conventional medical thinking by concluding that all this was evidence of a new type of disease. In 1982, he announced that he had purified and isolated the protein causing this disease. In 1997, he won the Nobel Prize in Physiology or Medicine for the discovery of prions.

A prion is a protein that has folded the wrong way. They spread not in the usual way of disease, but often by contact with other prions. When a prion comes in contact with a normal protein molecule, the prion somehow forces the normal protein to fold up incorrectly. Hence, the prionic disease can spread rapidly throughout the body.

Now, although there is still some disagreement, there are scientists who believe that many of the fatal diseases which afflict the elderly might also be caused by prions. Among them is Alzheimer's, which some have called the "disease of the century." Six million Americans are known to suffer from Alzheimer's, many of them sixty-five years or older. Fully one in three seniors die of Alzheimer's or dementia. Currently, it is the sixth-leading cause of death in the U.S., and cases are steadily rising. It is estimated that about half of people who survive into their eighties may eventually come down with this disease.

Alzheimer's is especially tragic because it strikes our most private and cherished possessions, our memories and sense of who we are. It first strikes the brain in areas near its center, like the hippocampus, which processes short-term memories. Therefore, the first signs of Alzheimer's disease are forgetting things that just happened. We may be able to recall events with razor-sharp accuracy that transpired sixty years ago, but forget things that happened six minutes ago. But eventually it attacks the entire brain, and even long-term memories disappear in the sands of time. It is always fatal.

My mother died of Alzheimer's disease. It was heartbreaking to see her memories slowly disappear, until she did not recognize who I was. Later, she did not know who she was.

Alzheimer's is known to have genetic links. People with a mutation in the APOE4 gene are more susceptible to the disease. In one BBC-TV series I hosted, the camera focused on my face as I was asked if I would take the APOE4 test to see if I was genetically prone to the disease. What would I say if I found out that I was indeed doomed to come down with Alzheimer's? I thought about it, and finally said I would still take the test, because it is always better to be prepared for the future, no matter what it holds. (Thankfully, my test was negative.)

Unfortunately, the root cause of Alzheimer's is unknown. The only way to confirm if someone had Alzheimer's is through an autopsy. Doctors often find that the brains of those with Alzheimer's have two types of sticky proteins, called the beta and tau amyloid proteins. But for decades, doctors have debated whether these sticky proteins are the cause of Alzheimer's or perhaps an unimportant by-product of the disease. The problem is that autopsies have shown that some people have had large amyloid deposits in their brains but were still totally free of any symptoms of the disease. So there is no direct cause-and-effect relationship between Alzheimer's and amyloid plaques in many cases.

One clue to this mystery was uncovered recently. Scientists in Germany found a direct correlation between people with misshapen proteins and those with Alzheimer's. In 2019, they made the stunning announcement that people with a misfolded amyloid protein in their blood, who were still free of symptoms, were twenty-three times more likely to get Alzheimer's disease. This link could even be confirmed up to fourteen years before a clinical diagnosis was made.

This means that perhaps years before you develop symptoms of Alzheimer's, a simple blood test might tell you the odds you will come down with it by searching for the misshapen amyloid protein.

Stanley Prusiner, in recent research he directed, has said, "I believe this shows beyond a shadow of a doubt that amyloid beta and tau are both prions, and that Alzheimer's disease is a double-prion disorder in which these rogue proteins together destroy the brain. . . . We need a sea change in Alzheimer's disease research."

One author of the report, Klaus Gerwert, stressed that this breakthrough could lead to new treatments for Alzheimer's, of which there are currently none: "The measurement of misfolded amyloid-beta in the blood may therefore make a key contribution toward finding a drug against Alzheimer's disease."

Hermann Brenner of Germany, another of the authors of the report, added, "Everyone is now pinning their hopes on using new treatment approaches during the symptom-free early stage of disease to take preventive steps."

"Good" and "Bad" Versions of Amyloid Protein

Yet another discovery made in 2021 might tell us precisely how this process occurs. Scientists at the University of California found that good and bad versions of the amyloid protein can be distinguished in one glance at their structure. They found that protein molecules, because they are made of a long string of amino acids which have curled up, often have clusters of atoms that spiral in one direction or the other, either clockwise or counterclockwise.

In the normal amyloid protein, the shape is "left-handed," i.e., the spirals and twists of the molecule are in one orientation. However, the other amyloid protein associated with Alzheimer's disease is right-handed. If this theory holds true, that one type of misshapen amyloid protein is responsible for Alzheimer's disease, it could represent an entirely new avenue for research.

First, we have to create detailed 3D images of these two types of amyloid proteins. Using quantum computers, it might be possible to see, at the atomic level, precisely how the misshapen Alzheimer's molecule is able to propagate by bumping into healthy molecules, and why it can cause so much damage to the brain.

Then, studying the structure of the protein, one might be able to determine how it derails the neurons in our nervous system. Once this mechanism is known, there are several possibilities. One way is to isolate the defects in this protein and use gene therapy to create a correct version of the gene. Or perhaps one day drugs may be devised that can either block the growth of the right-handed protein, or even help clear it out of the body more quickly.

For example, it is known that these misshapen molecules only exist in the brain for forty-eight or so hours before they are flushed naturally out. Once we understand the molecular structure of the right-handed protein, we can design another molecule that grabs this deviant molecule and then either breaks it up, neutralizes it so it is no longer dangerous, or binds with it so it is flushed out of the body more quickly. Quantum computers may be useful in finding its molecular weak spots.

In sum, quantum computers may identify many approaches at the molecular level that could neutralize or eliminate the bad prion, which we have been unable to do with trial and error and digital computers.

ALS

Yet another target for quantum computers is amyotrophic lateral sclerosis (ALS), also known as Lou Gehrig's disease, a fatal illness that reduces your body to a paralyzed mass of tissue and afflicts at least 16,000 people in the U.S. Your mind is intact, but your body wastes away. This disease attacks your nervous system, disconnecting your brain, in some sense, from your muscles, and eventually leading to death.

The most famous victim of the disease is the late cosmologist Stephen Hawking. His case was unusual, since he lived to be seventy-six, while most die rapidly. Victims of this dreaded disease usually live only two to five more years after being diagnosed.

Hawking once invited me to give a talk at Cambridge University on string theory. I was amazed when I visited his house. It was full of gadgets that allowed him to function in spite of this debilitating disease. In one mechanical device, you would place a physics journal. You would press a button, and then the device would grab a page and turn it for you automatically.

During the time that I had the pleasure of spending with him, I was deeply impressed by his willpower and his desire to be productive and participate in the physics community. Despite being almost totally paralyzed, he was determined to carry on his research and engage with the public. His determination in the face of monumental obstacles was a testament to his courage and drive.

Professionally, his work concerned the application of the quantum theory to Einstein's theory of gravity. The hope is that one day the quantum theory will return the favor and find a way for quantum computers to cure this horrible disease. At present, little is known about it because it is relatively rare. But by studying the family history of the victims, one can show that a series of genes are involved.

So far, about twenty genes have been found that are associated with ALS, but four of them account for most of the cases: C9orf72,

SOD1, FUS, and TARDBP. When these genes malfunction, they are associated with the death of motor neurons in the brainstem and spinal cord.

Of particular interest is the SOD1 gene.

It is believed that protein misfolding caused by SOD1 is implicated in ALS. The SOD1 gene makes an enzyme called superoxide dismutase, which breaks down charged oxygen molecules called superoxide radicals, which are potentially dangerous. But when SOD somehow fails to eliminate these superoxide radicals, nerve cells can be damaged. So the misfolding of the protein created by SOD1 could be one of the mechanisms that causes neurons to die.

Knowing the molecular pathway taken by these defective genes may be the key to curing the disease, and quantum computers could play a crucial role. Using the genes as a template, one can create a 3D version of the defective protein made by this gene. Then, studying the structure of the protein, one might be able to determine how it derails the neurons in our nervous system. If we can determine how the defective protein operates at the molecular level, we might be able to find a cure.

Parkinson's Disease

Yet another debilitating illness that involves mutated proteins in the brain is Parkinson's disease, which afflicts about one million people in the U.S. The most famous representative of this disease is Michael J. Fox, who has used his celebrity status to raise $1 billion to combat it. Typically it can cause one's limbs to tremble uncontrollably, but there are other symptoms as well, such as difficulty walking, loss of smell, and sleep disturbances.

There has been some progress with this disease. Scientists have found, for example, that, using brain scans, one can identify the precise location where neurons are overfiring and perhaps causing tremors in the hands. This form of Parkinson's can then be partially treated by inserting a needle in the brain where there is hyperactivity. Then, by neutralizing the neurons firing erratically, one can stop some of the tremors.

Unfortunately, there is still no cure. But some of the genes associated with Parkinson's have been isolated. It's possible to synthesize the proteins associated with these genes, whose 3D structure could be deciphered by quantum computers. In that way, we might discover how mutations in that gene can cause Parkinson's. We might be able to clone the correct version of that mutated protein and inject it back into the body.

So quantum computers may open up an entirely new way to approach these incurable diseases that afflict the elderly. Perhaps they can attack one of the greatest medical problems of all time: the aging process. If one can cure the aging process, then one simultaneously cures a host of diseases associated with it.

If quantum computers can one day find cures for the elderly, does it also mean that we won't have to die at all?

CHAPTER 13

IMMORTALITY

The oldest quest of all, stretching back to the earliest prehistory, is the search for immortality. No matter how powerful a king or emperor might be, they could never banish the wrinkles they saw in their reflection, foretelling their ultimate demise.

One of the earliest known tales, which predates parts of the Bible, is the Epic of Gilgamesh, the ancient Mesopotamian warrior, which chronicles his heroic exploits as he roamed across the ancient world. He engaged in numerous brave adventures as he rode the plains and deserts, even meeting a wise man who witnessed the Great Flood. Gilgamesh embarked on this journey because he was on a grand mission: to find the secret of living forever. At long last, he found the plant that was the source of immortality. But just before he could eat it, a snake suddenly snatched it from his hands and devoured it. Humans were not destined to become immortal.

In the Bible, God banished Adam and Eve from the Garden of Eden because they disobeyed His orders and ate the forbidden apple. But what was so dangerous about an innocent apple? It was because the apple was the forbidden fruit of knowledge.

Furthermore, God feared that by eating the apple from the tree of life, Adam and Eve would "become like one of us . . . and live forever"—they would become immortal.

The emperor Qin Shi Huang, the man who eventually united all of China around 200 BCE, was obsessed with the idea of immortality. In a famous legend, he sent his formidable naval fleet to search for the fabled Fountain of Youth. He gave his fleet one command: If you don't find the Fountain, then don't come back. Apparently, they did not discover the Fountain, but, banished from China, they went on to discover Korea and Japan instead.

According to Greek mythology, Eos, the goddess of the dawn, once fell in love with a mortal, Tithonus. Because mortals eventually die, Eos pleaded with the god Zeus to grant her lover immortality. Zeus granted her wish. But she made one decisive mistake. She forgot to ask for eternal youth for her lover as well. Sadly, each year Tithonus would get older and more and more decrepit, but could never die. So if one asks the gods for immortality, we must never forget to ask the gods to keep us forever young as well.

Today, armed with the advances of modern medicine, perhaps the time has come to revisit this age-old quest from a new perspective. By analyzing the mountain of genetic data on aging and teasing apart the molecular basis of life itself, one might use quantum computers to solve the problem of aging. In fact, quantum computers may be able to create two kinds of immortality, biological immortality and digital immortality. So the Fountain of Youth might not be a fountain after all, but a quantum computer program.

Second Law of Thermodynamics

Armed with modern physics, one can look back at this ancient quest from a modern perspective. The physics of aging can be explained using the laws of thermodynamics, that is, the laws of heat. There are three laws of thermodynamics. The First Law simply states that the total amount of matter and energy is a constant. You cannot get something from nothing. The Second Law says that in a closed system, chaos and decay always increase. The Third Law says that you can never reach absolute zero in temperature.

It is the Second Law that dominates our lives. It is a law of physics that mandates that things will eventually rust, disintegrate, and die. This means that entropy, which is a measurement of chaos, always increases. It seems that this iron law forbids immortality, because, in the end, everything falls apart. Physics seems to have a death warrant for all life on earth.

But there is a loophole in the Second Law. The fact that everything must decay only applies to a closed system. But in an open system, where energy can flow in from the external world, the increase of chaos can be reversed.

For example, every time a new life like a baby is born, entropy decreases. A new life represents a vast amount of data that is precisely assembled down to the molecular level. Life, therefore, seems to contradict the Second Law. But energy is flowing in from the outside in the form of sunlight. So energy from the sun is responsible for the creation of the vast diversity of life on the earth and the reversal of local entropy.

Thus immortality does not violate the laws of physics. There is nothing in the Second Law that forbids a life-form from living forever, as long as energy flows in from the outside. In our case that energy is sunlight.

What Is Aging?

So what is aging?

According to the Second Law, aging is primarily caused by the accumulation of errors at the molecular, genetic, and cellular level. Eventually, the Second Law catches up with us. Mistakes build up in our cells and DNA. Skin cells lose their elasticity and wrinkles form. Organs do not function properly and fail. Neurons misfire, so we forget things. Cancer sometimes develops. In short, we age and eventually die.

We can see this happening in the animal kingdom, which gives us clues to aging. Butterflies may live for a few days. Mice may live for a couple of years. But elephants may live for sixty to seventy years. And the Greenland shark may live for up to 500 years.

What is the common denominator here? Small animals lose heat rapidly compared to large animals. Hence, the metabolism rate of a mouse scurrying to avoid a predator is quite large, compared to a lumbering elephant leisurely eating a meal. But a higher metabolism rate also means a higher oxidation rate, which builds up errors in our organs.

Our car is a stark example of this. Where does aging take place in a car? Mostly, it takes place in the engine, where we have oxidation due to the burning of fuel, and also the wear and tear of moving gears. But where is the engine of the cell?

Most of the energy of the cell originates in the mitochondria. We suspect, therefore, that the mitochondria is where much of the damage from aging accumulates. It is likely that aging might be reversed if we evade the Second Law by adding energy from the outside, in the form of a better, healthier lifestyle, and also genetic engineering to fix broken genes.

Now, think of a car filled with high-octane fuel. The car runs beautifully. Even an aging car can run better with supercharged gasoline. This, in turn, is similar to what hormones like estrogen and testosterone do to the human body. In some sense, they act like the elixir of life, giving us energy and vitality beyond our years. Some believe that estrogen is one reason why women on average live longer than men. But there is a price we have to pay for this extra mileage. And that is cancer. Extra wear and tear also means more errors build up, among them the genes for cancer. So in a sense cancer represents the Second Law of Thermodynamics catching up with us.

These errors in our DNA happen all the time. DNA lesions at the molecular level, for example, happen 25 to 115 times per minute in our body, or about 36,000 to 160,000 per cell per day. We also have a DNA repair mechanism in our body, but aging accelerates when these repair mechanisms are overwhelmed by the sheer number of errors in our DNA. Aging occurs when the buildup of errors exceeds our ability to repair them.

Predicting How Long We Can Live

If aging is related to errors in our DNA and cells, then it might be possible to get a rough numerical principle that predicts how long we can live.

One intriguing study was done by the Wellcome Sanger Institute in Cambridge, England. If aging is related to genetic damage, then one can predict that the more damage an animal has, the shorter its life span. Sure enough, these Cambridge scientists found an inverse relationship after analyzing sixteen species of animals: the more the genetic damage, the shorter the life span.

They found remarkable correlations between animals that are quite dissimilar. The tiny naked mole rat suffers 93 mutations per year and can live to twenty-five to thirty years. Meanwhile, the giant giraffe can suffer 99 mutations per year over a twenty-four-year life span. If we multiply these two numbers, the product is roughly 2,325 total mutations for the mole and 2,376 for the giraffe, which are notably similar. Although these two mammals differ in significant ways, they accumulate roughly the same number of mutations in their lifetime.

This gives us a formula that can roughly predict the life span of humans by analyzing data from many animals. When analyzing mice, they found that they had 793 mutations per year, spread out over a lifespan of 3.7 years, for a total of 2,934.1 total mutations.

The numbers for humans are a bit trickier, since they vary across different cultures and locations. Humans, it is believed, have 47 mutations per year. Most mammals, on average, have 3,200 mutations across a lifetime. It would mean that, as a first guess, humans have a life span of about seventy. (With a different set of assumptions, one can also arrive at a figure of about eighty years.)

The results of this simple calculation are rather remarkable. They indicate the importance of genetic errors in our DNA and cells as one of the main drivers of aging and eventual death.

So far, all these results were done on animals in the wild, in their natural state. But what happens when we subject animals to different external conditions? Is it possible to change their life span artificially?

The answer seems to be yes.

Resetting the Biological Clock

With medical intervention (e.g., genetic engineering, changes in lifestyle) it might be possible to extend the human life span by correcting the damage caused by the Second Law.

There are several possibilities. One possibility is to reset the "biological clock." When a cell reproduces itself, the chromosomes get slightly shorter. For skin cells, after about sixty reproductions, the cell starts to age, which is called senescence, and eventually dies. This number is known as the Hayflick limit. It is one reason why cells die, because they have a built-in clock that tells them when to do so.

I once interviewed Leonard Hayflick about his famous limit. He was cautious, however, that some people might jump to too many conclusions about this biological clock. We are just beginning, he told me, to understand the aging process. He bemoaned the fact that the field of biogerontology, the science of aging, had to deal with so much misinformation in the public, especially the latest dietary fad.

The Hayflick limit occurs because there is a cap, called a telomere, at the end of the chromosome, which gets shorter with each reproduction. But like the tips of your shoelace, after too many manipulations, the cap wears off and the shoelace starts to fray. After sixty or so reproductions, the telomeres wear off, the chromosome frays, and the cell goes into senescence and eventually dies.

But it is also possible to "stop the clock." There is an enzyme, called telomerase, which can prevent the telomeres from becoming increasingly short. At first glance, one might think that this could be the cure for aging. In fact, scientists have been able to apply telomerase to human skin cells, so that they have divided hundreds of times, not just sixty. This research has allowed us to "immortalize" at least one life-form.

But there are also dangers involved. It turns out that cancer cells also use telomerase to attain immortality. In fact, the presence of telomerase has been detected in 90 percent of all human tumors. One must be careful when manipulating telomeres in the body so we don't accidentally convert healthy cells into cancerous ones.

So if we ever find the Fountain of Youth, telomerase may be part of the solution, but only if we can cure its side effects. Quantum computers may be able to solve the mystery of how telomerase can cause a cell to become immortal but not cancerous. Once this molecular mechanism is found, it might be possible to modify the cell so that it has an extended life span.

Caloric Restriction

In spite of all the quack cures and therapies over the centuries to increase our life span, one method has withstood the test of time and seems to work in every case. The only proven way in which to lengthen the life span of an animal is through caloric restriction. In other words, if you eat 30 percent fewer calories, you can live roughly 30 percent longer, depending on the animal being studied. This general rule has been tested across a vast array of species, from insects, mice, dogs and cats, even to apes. Animals eating fewer calories live longer than their counterparts that gorged themselves. They have fewer diseases and suffer less frequently from the problems of old age, such as cancer and hardening of the arteries.

Although this has been tested among animals across the animal kingdom, one, however, hasn't been systematically analyzed in this way so far: *Homo sapiens*. (This is probably because we live too long, and would complain about eating a spartan diet so meager it would make a hermit hungry.) No one knows precisely why this works, but one theory posits that eating less reduces your oxidation rate, thereby slowing the aging process.

One experimental result that seems to vindicate this theory can be found with worms like *C. elegans*. When these worms are genetically altered to reduce their oxidation rate, their life span can be extended many times. In fact, scientists have given names to some of these genes like Age-1 and Age-2. Reducing the oxidation rate seems to help the cells repair damage. So it seems reasonable that caloric restriction works by decreasing the oxidation rate in our body, which decreases the accumulation of errors.

But this leaves open one question: Why do some animals exhibit caloric restriction in the first place? Do animals consciously eat less to live longer? (One theory states that animals, in their natural state, have two choices. On the one hand, they can reproduce and have young. But this requires a steady, plentiful food supply, which is rare. More common is that most animals are near starvation, constantly hunting and scavenging for food. So in times of scarcity, which is more often than not the case, animals have evolved to instinctively eat less, in order to conserve energy and live longer, until the time comes when food is plentiful and they can reproduce.)

Scientists who have studied caloric restriction believe that it may work via the chemical resveratrol, which in turn is produced by the sirtuin gene. Resveratrol is found in red wine. (This has created a mini-fad around resveratrol and red wine, but the jury is still out on whether resveratrol can truly extend the human life span.)

But in 2022, studies done at Yale University may finally have solved part of the riddle of why caloric restriction actually works. They concentrated their efforts on the thymus gland, located between the lungs, which manufactures T cells, an important player among our white blood cells, which help defend against disease. They noticed that these T cells from the thymus gland age faster than ordinary T cells. When we hit the age of forty, for example, 70 percent of the thymus is fatty and nonfunctional. Vishwa Deep Dixit, a senior author of the paper, says, "As we get older, we begin to feel the absence of new T cells because the ones we have left aren't great at fighting new pathogens. That's one of the reasons why elderly people are at greater risk for illness." If true, this may explain why the elderly are more prone to age and die.

Given this result, they performed another experiment that involved putting one group of people on a calorie-restricted diet for two years. They were surprised to find that this group had less fat and more functioning cells in their thymus gland. This was a remarkable result.

Dixit adds, "The fact that this organ can be rejuvenated is, in my view, stunning because there is very little evidence of that happening in humans. That this is even possible is very exciting."

The Yale group began to realize that they were on to something important. Next, they had to investigate the root cause: How, at the molecular level, does caloric restriction boost the immune system?

They were eventually able to zero in on the protein called PLA2G7, which is involved with inflammation, another phenomenon associated with aging. "These findings demonstrate that PLA2G7 is one of the drivers of the effects of caloric restriction. Identifying these drivers helps us understand how the metabolic system and the immune system talk to each other, which can point us to potential targets that can improve immune function, reduce inflammation, and potentially even enhance healthy lifespan," says Dixit.

The next step would be to use quantum computers to find out

how at the molecular level this protein can reduce inflammation and retard the aging process. Once this process is understood, it may be possible to manipulate PLA2G7 and reap the benefits of caloric restriction without having to put ourselves on a starvation diet.

Dixit concludes by saying that his study into the relevant proteins and genes could change the direction of research on the aging process. He concludes, "I think it gives hope."

Key to Aging: DNA Repair

But this raises another question: How does caloric restriction repair the molecular damage caused by oxidation? Caloric restriction may work by slowing down the oxidation process, making it possible for the body to repair the damage it causes naturally, but how does the body repair DNA damage in the first place?

This is being studied at the University of Rochester, where scientists are investigating whether the DNA repair mechanism can be understood by examining the animal kingdom. More specifically, can DNA repair mechanisms explain why some animals live longer? Is there a genetic Fountain of Youth?

They analyzed the life span of eighteen species of rodents, and found something interesting. Mice may live for just two to three years, but beavers and naked mole rats can live to the astonishingly ripe old age of twenty-five to thirty years. Their theory is that the long-lived rodents have a stronger DNA repair mechanism than the short-lived rodents.

To investigate this, they focused on the sirtuin-6 gene, which is involved with DNA repair, and is sometimes dubbed the "longevity gene." They discovered that not all sirtuin-6 proteins are the same. There are five different types of proteins created by sirtuin-6, and each had different levels of activity. They also noted that beavers had sirtuin-6 proteins which were more potent than the proteins created by rats, though not naked mole rats. This, they claimed, might be the reason why the beavers lived so long.

To prove their theory, they injected the various sirtuin-6 proteins into different animals to see if it affected their life span. Fruit flies injected with the beaver sirtuin-6 protein lived longer than fruit flies injected with the rat protein.

When injected into human cells, they found a similar effect. Cells that received the beaver sirtuin-6 protein sustained less DNA damage than cells that had the rat protein. Vera Gorbunova, one of the researchers, says, "If diseases happen because of DNA that becomes disorganized with age, we can use research like this to target interventions that can delay cancer and other degenerative diseases."

This is important, because repairing DNA damage that may be regulated by genes like sirtuin-6 could be the key to reversing the aging process. Quantum computers can then be used to determine precisely how sirtuin-6 is able to enhance DNA repair mechanisms at the molecular level.

Once this process is unraveled, it may be possible to find ways to accelerate it, or to find new molecular pathways that can stimulate DNA repair mechanisms. So if DNA damage is one of the drivers of the aging process, then it is crucial to understand how it can be reversed at the molecular level using quantum computers.

Reprogramming Cells for Youth

However, the danger is that there is a lot of quackery when it comes to trying to live longer. There is always the fad-of-the-month: the latest vitamin, herb, or "miracle cure." But there is one serious organization that has been getting a lot of publicity concerning the aging process.

Russian billionaire Yuri Milner, who made his fortune on Facebook and Mail.ru, has assembled a blue-ribbon group of top academics to look into the question of reversing aging. He is a well-known figure in Silicon Valley, giving $3 million per year for his Breakthrough Prize to outstanding physicists, biologists, and mathematicians.

Now, his attention is focused on a new group called Altos Labs, which wants to use the science of "reprogramming" to perhaps rejuvenate aging cells. Even Jeff Bezos is among the deep-pocketed investors lining up to back Altos. According to a paper filed by Altos, the fledgling company has already lined up $270 million.

According to MIT's *Technology Review,* the idea behind this effort is reprogramming the DNA of aging cells so that they revert back to an earlier form. This was experimentally tested by Japanese Nobel laureate Shinya Yamanaka, who will chair Altos's scientific advisory board.

Yamanaka is one of the world's authorities on stem cells, which are the mother of all cells. Embryonic stem cells have the remarkable property of being able to turn into any cell of the human body. What Yamanaka found was a way to reprogram adult cells so that they reverted back to their embryonic state, so that they could, in principle, create entirely new, fresh organs from scratch.

The key question is: Can you reprogram an aging cell to become youthful again? What is driving the interest in Altos is that the answer is apparently yes—under certain circumstances, there are four genes (now called Yamanaka factors) that can perform the reprogramming process.

In some sense, reprogramming of aging cells is commonplace. Think of how Mother Nature can take the cells of an adult and reprogram them to become the stem cells of an embryo. So reprogramming is not science fiction; it is a fact of life. This rejuvenation process happens in every generation, when the embryo is first conceived.

Not surprisingly, a number of start-ups, always looking for the next big thing, have jumped onto this bandwagon, including Life Biosciences, Turn Biotechnologies, AgeX Therapeutics, and Shift Bioscience. "If you see something in the distance that looks like a giant pile of gold, then you should run quick," says Martin Borch Jensen, of Gordian Biotechnology. In fact, he is giving out $20 million to speed up research.

David Sinclair, a professor at Harvard, has said, "There are hundreds of millions of dollars being raised by investors to invest in reprogramming, specifically aimed at rejuvenating parts or all of the human body." Sinclair was able to use this technique of reprogramming cells to restore eyesight in mice. He adds, "In my lab, we are ticking off the major organs and tissues, for instance skin, muscle and brain, to see which we can rejuvenate."

Alejandro Ocampo of the University of Lausanne in Switzerland says, "You can take a cell from an 80-year-old, and, in vitro, reverse the age by 40 years. There is no other technology that can do that."

An independent group at the University of Wisconsin–Madison took samples of synovial fluid (which is a thick liquid found in the joints of your body), which contains certain stem cells called MSCs (mesenchymal stem/stromal cells). It was previously known that it is possible to reprogram MSC cells so they become younger. But the way in which this rejuvenation takes place is unknown.

They were able to fill in many of the missing steps. MSC cells were converted into induced pluripotent stem cells (called iPSCs) and converted back into MSC cells. After this round-trip, they found that the reprocessed MSC cells were rejuvenated. Most important, they were able to identify the specific chemical pathway that took MSC cells on this round-trip. Involved in this process was a series of proteins and genes called GATA6, SHH, and FOXP.

These are remarkable breakthroughs that were once thought to be impossible. So scientists are beginning to understand how aging cells can become youthful once again.

But there is also reason to be cautious. We saw previously that methods to delay or reverse aging included side effects such as cancer. Estrogen can keep women fertile for many years until menopause, but cancer is one possible side effect of this hormone. Similarly, telomerase can stop the clock on cell aging, but also introduces increased cancer risk.

Similarly, one of the dangers of reprogramming cells is cancer. Research has to proceed carefully so that dangerous side effects do not sidetrack it. Quantum computers may prove helpful in this effort. First, they may be able to unravel the rejuvenation process at the molecular level and find the secrets behind embryonic stem cells. Second, it may be possible to control some of the side effects of this process, such as cancer.

Human Body Shop

Yet another experiment has piqued interest in cell rejuvenation.

In the original Yamanaka approach, skin cells were exposed to the four Yamanaka factors for fifty days in order for them to revert back to an embryonic state. But scientists at the Babraham Institute in Cambridge, England, exposed these cells for only thirteen days, and then let them grow normally.

The original skin cells were taken from a fifty-three-year-old woman. The scientists were shocked to find that the rejuvenated skin cells looked and acted like they were from a twenty-three-year-old.

"I remember the day I got the results back and I didn't quite believe that some of the cells were 30 years younger than they were supposed to be. . . . It was a very exciting day," said Diljeet Gill, one of the scientists conducting the study.

The results were sensational. If this result is verified, then apparently it represents the only time in medical history that scientists have successfully rejuvenated aging cells, so they behaved as if they were decades younger.

However, the scientists involved in the study were careful to mention possible side effects. Because of the enormous genetic changes involved with rejuvenation, as is true for so many promising treatments, cancer remains a possible by-product. So this entire approach has to proceed carefully.

But there is a second way to create youthful organs, without the danger of cancer: tissue engineering, where scientists literally build human parts from scratch.

Tissue Engineering

If an adult cell reverts back to an embryonic state, it does rejuvenate, but it does this only at the cellular level. This means you cannot rejuvenate the entire body and live forever. It only means that certain cell lines become immortal, so that specific organs may be regenerated, but not the entire body.

One reason for this is that stem cells, if left to themselves, sometimes create a formless mass of random tissue. Stem cells often need cues from neighboring cells in order to grow correctly in sequence, to create the final organ.

The solution to this may be tissue engineering, which means putting stem cells in a mold of some sort so that the cells grow in an orderly fashion.

This approach has been pioneered by Anthony Atala, of Wake Forest University in North Carolina, and others. I had the honor of interviewing Atala for BBC-TV. As I walked around his laboratory, I was amazed to see large jars containing human organs, like livers, kidneys, and hearts. I almost felt like I had stepped into a science fiction movie.

I asked him how his research is performed. He told me that he first creates a special mold made of tiny plastic fibers in the shape of the organ he wants to grow. Then he seeds the mold with cells of the organ taken from the patient. Next, he applies a cocktail of growth factors to stimulate these cells. The cells begin to grow into the fibers of the mold. Eventually, the mold, which is biodegradable, disappears and leaves behind a near-perfect copy of the organ. Then the artificial organ is placed inside a patient's body, where it starts to function. Because the cells are made from the patient's own tissue, there is no rejection mechanism, which is one of the main problems facing organ transplants. There is also no danger of cancer, since he is not manipulating the delicate genetics inside a cell.

He told me that most of the organs that have been successfully made consist of just a few cell types. This includes skin, bone, cartilage, blood vessels, bladders, heart valves, and windpipes. The liver is more difficult, he said, because it consists of several different types of cells. And the kidney, because it consists of hundreds of tiny tubes and filters, is still an ongoing project.

His approach may also be combined with stem cells, so that it may one day be possible to regenerate entire organs of the body as they wear out. For example, since cardiovascular diseases are the number one cause of death in the U.S., it may be possible one day to grow an entire heart in the laboratory. It would be like creating a "human body shop."

Other groups are experimenting with 3D printing to create human organs. In the same way that a computer printer can shoot out tiny droplets of ink to form an image, it can be modified to shoot out individual human heart cells, to create heart tissue cell by cell. If cell rejuvenation is successful in creating youthful cell lines, then tissue engineering may be able to grow any organ of the body using stem cells, such as the heart.

In this way, we avoid the problem faced by Tithonus.

Role of Quantum Computers

Quantum computers may have a direct impact on these endeavors. In the near future, most of the human population will have their genome sequenced and included in a giant global gene bank. This huge storehouse of genetic information may overwhelm a conventional digital computer, but analyzing incredible amounts of data is precisely what quantum computers excel at. This may allow scientists to isolate the genes affected by the aging process.

For example, scientists can already analyze the genes of young people and the elderly and compare them. In this way, about 100 or so genes where aging seems to be concentrated have been identified. Many of these genes, it turns out, are involved in the oxidation process. In the future, quantum computers will analyze an even larger mass of genetic data. This will help us understand where most genetic and cellular errors accumulate, but also which genes might actually control aspects of the aging process.

Quantum computers may not only isolate the genes where most of the aging takes place, they may also do the opposite: isolate the genes that are found in exceptionally old but healthy people. Demographers know that there are the super-aged, i.e., individuals who seem to have beaten the odds and live a healthy robust life much longer than expected. So quantum computers, analyzing this mass of raw data, may find the genes that indicate an exceptionally healthy immune system and allow the elderly to reach a ripe old age by avoiding the diseases that might strike them down.

Of course, there are also individuals who age so rapidly that they die of old age as children. Diseases like Werner syndrome and progeria are a nightmare, where children age almost before your eyes. They rarely live beyond their twenties or thirties. Studies have shown that, among other problems, they have short telomeres, which may partly contribute to their accelerated aging. (By the same token, studies on Ashkenazi Jews have found the opposite, that long-lived subjects had a hyperactive version of telomerase, which may explain their long life.)

Furthermore, tests of people older than one hundred show that they have a significantly higher level of the DNA repair protein called poly (ADP-ribose) polymerase (PARP) than younger individuals aged twenty to seventy. This indicates that longer-lived individuals have stronger DNA repair mechanisms to reverse genetic damage and hence live longer. These centenarians also have cells that resemble cells taken from much younger people, indicating that aging has slowed. This, in turn, may explain the curious fact that those who reach their eighties have a greater-than-normal chance of living into their nineties and beyond. This may be because people with weak immune systems die before they reach their eighties, so the ones who survive have stronger DNA repair mechanisms, which can prolong their life span into their nineties and beyond.

So quantum computers may be able to isolate key genes in several categories:

- The elderly who are exceptionally healthy for their age
- Individuals who have immune systems that can fight off common diseases, thereby prolonging life
- Individuals who have accumulated errors in their genes that have accelerated aging
- Individuals who have deviated significantly from the norm, such as those who have aged extraordinarily fast from diseases like Werner syndrome and progeria

Once the genes associated with aging are isolated, then perhaps CRISPR may be able to fix many of them. The goal is to fix the genes where most aging takes place, using quantum computers to isolate the precise molecular mechanisms of that process.

In the future, perhaps a cocktail of different drugs and therapies will be developed that can slow down and possibly reverse aging. The combined effect of different medical interventions acting in concert may be able to roll back the hands of time.

The key is that quantum computers will be able to attack the aging process in the arena in which it takes place: at the molecular level.

Digital Immortality

In addition to biological immortality, there is the real possibility that we might achieve digital immortality using quantum computers.

Most of our ancestors lived and died without leaving a trace of their existence. Perhaps there is a line in church or temple records, documenting when our ancestors were born, and a second line documenting when they died. Or perhaps there is a broken tombstone in a deserted graveyard with our ancestor's name on it.

And nothing more.

An entire lifetime of cherished memories and experiences has been reduced to two lines in a book and some engraved stone. People who use DNA to trace their lineage often find out that very quickly their trail disappears within a century. Their entire family history is reduced to dust after a generation or two.

But today, we leave a formidable digital footprint. Our credit card transactions alone can give a reasonable glimpse into our history and personality, our likes and dislikes. Every purchase, vacation, sporting event, or gift is recorded in some computer. Without even realizing it, our digital footprint creates a mirror image of who we are. In the future, this mass of information can give us a digital re-creation of our personality.

Already, people are talking about resurrecting historical figures and well-known individuals through a digitization process that will make them available to the public. Today, you may go to the library to look up a biography of Winston Churchill. In the future, you may instead talk to him. All his letters, memoirs, biographies, interviews, etc. will be digitized and made available. You might talk to a holographic image of the former prime minister and spend a leisurely afternoon engaging in a revealing conversation with the man.

I personally would love to spend time talking to Einstein, to ask him about his goals, his achievements, and his philosophy of science. What would he think, realizing that his theories have blossomed into huge scientific disciplines like the Big Bang, black holes, gravity waves, the unified field theory, among others? What would he think of how the quantum theory has evolved with time? He left an extraordinarily large collection of letters and personal correspondence that reveal his true character and thoughts.

Eventually, the average person may achieve digital immortality too. In 2021, William Shatner, the star of the *Star Trek* TV series, attained a form of digital immortality. He was placed before a camera and for four days was asked hundreds of personal questions, about his life, his goals, his philosophy. Then a computer program analyzed this mass of material and arranged them chronologically, according to subject matter, place, etc. In the future, you may be able to ask personal questions directly to this digitized Shatner, and it will answer back in a coherent, rational way, as if he were there talking to you in your living room.

In the future, you will not need to sit in front of a TV camera to be digitized. Unconsciously, without thinking about it, we use the camera in our cell phone to record our daily activities and lives. In fact, many teenagers already create a huge digital footprint as they document their pranks, jokes, and antics (some of which may live forever on the internet).

Normally, we think of our lives as a series of accidents, coincidences, and random experiences. But with enhanced AI, we will one day be able to edit this treasure trove of memories and arrange it in an orderly fashion. And quantum computers will help sort through this material, using search engines to find missing background material and editing the narrative.

In some sense, our digital selves will never die.

So perhaps our legacy of cherished personal memories and achievements does not have to dissipate and scatter with the shifting sands of time when we pass away. Perhaps quantum computers will give us a form of immortality.

In summary, scientists are now beginning to identify some of the pathways involved in extending the human life span. It is still a mystery, however, how these pathways actually work at the molecular level. For example, how can certain proteins accelerate the molecular repair of DNA? Quantum computers may play a decisive role, because only a quantum system can fully explain another quantum system like molecular interactions. Once the precise mechanisms of things like DNA repair are known, one might be able to improve on it to delay or even arrest the aging process.

Quantum computers may also give us the ability to live forever digitally. When combined with artificial intelligence, we should be able to create a digital copy of ourselves that accurately reflects who we are. Steps are already being taken to perfect this process.

But the next frontier for quantum computers is not just the application of quantum mechanics to the inner space of our bodies, but to apply quantum computers to the external world, solving pressing problems such as global warming, harnessing the power of the sun, and deciphering the mysteries of the world around us. The next goal is to use quantum computers to understand the universe.

PART IV

MODELING THE WORLD AND THE UNIVERSE

CHAPTER 14

GLOBAL WARMING

I once gave a lecture at the university in Reykjavík, the capital of Iceland.

As the plane approached the airport, I looked out at the barren volcanic landscape that was almost devoid of vegetation. It seemed like a journey back in time. The area near the airport was so desolate, it made the perfect place to see back millions of years into the past.

Later, I was given a guided tour of the campus and was keen to see their research on ice cores, which can chronicle the weather over thousands of years.

Their laboratory was in a large room that resembled a huge freezer and was as cold as one. I noticed that there were several long metal rods laid out on a table. The rods were about 1.5 inches in diameter and many feet long, and each contained a core sample taken from deep in the ice.

Some of the rods were open, and you could see that they contained long cylinders of white ice. I shuddered when I realized that I was looking at ice that fell onto the Arctic thousands of years ago. I was staring into a time capsule from well before recorded history.

Looking carefully at these ice cores, I could see a series of thin, brown horizontal bands along the ice. I was told by the scientists that each band was created by the soot and ash released by ancient volcanic eruptions.

By measuring the spacing between the various bands, you could determine their age by comparing them to known volcanic eruptions.

They also told me that within the ice cores, there are microscopic air bubbles which are like a snapshot of the atmosphere thousands of years ago. By determining their chemical content, one can easily determine the amount of CO_2 that existed back then.

(Calculating the temperature when the ice cores formed is more difficult, and is done indirectly. Water consists of hydrogen and oxygen as H_2O. But there is a heavy version of water, where the O-16 and H-1 atoms are replaced by an isotope with extra neutrons, creating O-18 and H-2. The heavier version of H_2O evaporates more quickly when it is relatively warm. Thus, by measuring the ratio between the heavy water molecules and the normal molecule, one can calculate the temperature when the ice first formed. The more heavy water there is, the colder it was when the snow first fell.)

Finally, I saw the results of their painstaking but revealing work. On a graph, the temperature and CO_2 content over the centuries were like a pair of roller coasters, going up and down in unison. Clearly, there was a tight, important correlation between the temperature of the planet and the CO_2 content in the air. (Today, these ice cores can go back even further. In 2017, scientists were able to extract ice cores in Antarctica that were 2.7 million years old, giving scientists a previously unknown history of our own planet.)

Several things impressed me while analyzing this chart. First, you notice wild swings in temperature. We think of the earth as being so stable. Yet we are reminded that it is a dynamic object, with large gyrations in the temperature and climate.

Second, you notice that the last ice age ended around 10,000 years ago, when much of North America was buried under almost half a mile of solid ice. But since then, there has been a gradual heating of the atmosphere, which made possible the rise of human civilization. Since we will probably have another ice age in 10,000 years or so, it means that the rise of human civilization accidentally took place because we entered an interglacial period between two ice ages. Without this thaw, we would still be living in small, nomadic bands of hunters and scavengers, wandering in the ice and desperately looking for scraps of food.

But what caught my eye was that there has been a slow rise in temperature since the last ice age ended 10,000 years ago, but then a sudden spike in temperature within the last 100 years, coinciding with the coming of the Industrial Revolution and the burning of fossil fuels.

In fact, by analyzing temperatures around the planet, scientists concluded that the years 2016 and 2020 went down as the hottest years ever recorded in history. In fact, the period from 1983 to 2012 was the hottest thirty-year period in the last 1,400 years. So the recent heating of the earth is not a by-product of the warming due to the interglacial period, but something highly unnatural. The leading candidate, among many factors, for this is the rise of human civilization.

Our future may hinge on our ability to predict weather patterns and plot realistic courses of action. We are now pushing the limit of what conventional computers can perform, so we will need to turn to quantum computers to give us an accurate assessment of global warming and "virtual weather reports" of possible futures, allowing us to vary certain parameters to see how they affect the climate.

One of these virtual weather reports may hold the key to the future of human civilization.

As Ali El Kaafarani writes in *Forbes* magazine, "Quantum computers also hold immense potential from an environmental perspective, and experts predict that, through quantum simulations, they will be instrumental in helping countries meet the United Nation's Sustainable Development Goals."

CO₂ and Global Warming

Most of all, we need accurate assessments of the greenhouse effect, and how human activity is contributing to it.

Light from the sun can easily penetrate the earth's atmosphere. But when it is reflected off the surface of the earth, it loses energy and becomes infrared heat radiation. But because infrared radiation does not penetrate CO_2 very well, the heat is trapped on the earth, thereby heating it up. Eighty percent of the world's energy in 2018 came from the burning of fossil fuels, which produces CO_2 as a by-product. So the sudden spike in temperature within the last century is probably caused by a number of factors, especially the buildup of CO_2 as a result of the Industrial Revolution.

The rapid heating of the earth in the last 100 years has also been confirmed from an entirely different source, not from the inner space of underground ice cores, but from outer space. From that vantage point, the effects of global warming are quite visually dramatic.

For example, NASA weather satellites can calculate the total amount of energy that the earth receives from the sun. These satellites can also determine the total amount of energy that the earth sends back into outer space. If the earth was in equilibrium, we would see that the input and output of energy are roughly the same. When all factors are carefully considered, one finds that the earth absorbs more energy than it radiates back into space, causing the earth to heat up. If we then compare the net amount of energy captured by the earth, it is about the same as the amount of energy generated by human activity. So the main culprit driving the recent increase in heating the planet seems to be human activity.

Satellite photos reveal the consequences of this warming. These photos today can be compared to photos taken decades ago, showing the stark changes in the earth's geology. We see that all the major glaciers have receded over the decades.

Submarines have visited the North Pole since the 1950s. They have determined that the ice during the winter months has become thinner by 50 percent in the last fifty years, decreasing in thickness by about 1 percent per year. (Children in the future may wonder why their parents talk about Santa Claus being from the North Pole, when there is almost no more polar ice at all.) According to NASA scientists, by mid-century, the Arctic Ocean will be completely ice-free in the summer.

Hurricane activity may also change. They start as a mild tropical wind off the coast of Africa and then migrate across the Atlantic Ocean. Once they hit the Caribbean, they are like bowling balls. If they hit at just the right angle, they can enter the warm waters of the Gulf of Mexico and then grow in intensity to become monster storms. The intensity, frequency, and duration of hurricanes hitting the East Coast have all increased since the 1980s, probably due to increases in water temperature. Hence we will probably see hurricanes of increasing power and devastation in the future.

Future Predictions

The computer projections for the future of the earth's climate are quite bleak. Global sea levels have increased by eight inches since 1880. (This is because the temperature of the oceans is increasing, which causes the total volume of ocean water to expand.) Most likely, it will rise one to eight feet by 2100. Maps of the world in 2050 to 2100 show a striking change of the coastal areas.

"Sea-level rise driven by global climate change is a clear and present risk to the United States today and for the coming decades and centuries," states a report from NASA and NOAA, the National Oceanic and Atmospheric Administration.

But for every inch you lose vertically, coastal areas can lose 100 inches horizontally in terms of usable coastline. So the very map of the earth is gradually changing. Furthermore, sea levels will continue to rise well into the twenty-second century because of the enormous amount of heat already circulating in the atmosphere. At the very least, this means that coastal areas will experience large-scale flooding as ocean waves begin to surge past dams and barriers.

Bill Nelson, NASA administrator, comments on the recent NASA/NOAA report on the weather: "This report supports previous studies and confirms what we have long known: Sea levels are continuing to rise at an alarming rate, endangering communities around the world. . . . Urgent action is required to mitigate a climate crisis that is well underway."

Coastal cities around the world will have to deal with the rising water. Venice is already underwater during certain times of the year. Parts of New Orleans are already below sea level. All coastal cities will need to have plans to accommodate the rise of the sea level into the coming decades, such as locks, levies, dikes, evacuation zones, hurricane warning systems, and so forth.

Methane as a Greenhouse Gas

Methane is actually over thirty times more potent than carbon dioxide as a greenhouse gas. The danger is that the Arctic regions near Canada and Russia, which contain vast stretches of tundra, may be thawing out, releasing methane gas.

I once gave a lecture in Krasnoyarsk, in Siberia. The residents there told me they actually did not mind global warming, since it meant that their homes were not continually frozen in. They also told me a curious fact: the huge carcasses of mammoths that died tens of thousands of years ago are emerging from the ice as temperatures rise.

Although the locals who live in Siberia might not mind the milder weather, the real danger is to the rest of the globe, where the release of methane gas may cause a runaway ripple effect. The more the earth heats up, the more the tundra melts and releases methane gas. But this methane in turn heats up the earth even more and starts the cycle all over again. So the more the tundra melts, the more our planet warms. Since methane is a potent greenhouse gas, this means that many computer projections in the future may actually underestimate the true magnitude of global warming.

Military Implications

We see the effects of global warming everywhere. Farmers, for example, are in tune with the cycles of the weather, and they are well aware that the summers are about a week longer on average than they used to be. This affects when they plant seeds and what plants they raise that year.

Insects, like mosquitoes, are also moving northward, perhaps bringing tropical diseases with them, like the West Nile virus.

Because the energy circulating in the weather is increasing, it means more violent swings in the weather, and not just a steady increase in temperature. Thus, we can expect forest fires, droughts, and floods to become more and more common. "One-hundred-year storms" once described very rare but violent events, but now they seem to be occurring with greater frequency. In 2022, Europe and the U.S. were hit with especially high temperatures that broke records across much of the planet, creating massive forest fires, disappearing lakes, and deaths due to dehydration, among other serious consequences.

Ominously, the poles, which exert an enormous amount of influence on the weather, have heated up faster than other regions of the planet. The amount of melting in Greenland just within the last twenty years has created enough liquid water to cover all of the United States with 1.5 feet of water.

Meanwhile, Antarctic ice sheets have developed underground rivers of freshly melted snow. It now seems clear that the poles are not as stable as previously thought.

A recent NASA/NOAA report focused on the possible collapse

of the Thwaites Glacier in Antarctica, which has been nicknamed the "doomsday glacier." "The eastern ice shelf is likely to shatter into hundreds of icebergs. Suddenly, the whole thing would collapse," says Erin Pettit, a glaciologist from Oregon State University.

This also has geopolitical and military implications. The Pentagon once drafted a worst-case scenario if global warming grows out of control. It identified one of the deadliest hotspots as the border between Bangladesh and India. Because of sea level rise and intense flooding, global warming may one day force millions of people from Bangladesh to flee and rush the border with India. This mass of desperate people could easily overwhelm border guards. Then there would be mounting pressure on the Indian military to beat back wave upon wave of refugees trying to escape the floodwaters. As a last resort, the Indian military might be asked to protect its borders by using nuclear weapons.

This was a worst-case scenario, but it graphically illustrates what might happen if things spiral out of control.

Polar Vortex

Some people point to recent monster snowstorms that have engulfed huge portions of the U.S., and assert that the threat of global warming is highly exaggerated.

But one has to look at the reason for this instability in the winter weather. Whenever there is a huge winter storm, the weather report details the motion of the jet stream as it meanders its way down from Alaska and Canada, bringing freezing weather with it.

The jet stream, in turn, follows the gyrations of the polar vortex, a narrow, spinning cylinder of super-cold air that is centered on the North Pole. Recently, satellite photographs of the polar vortex show that it is becoming more unstable, so that it wanders, sending the jet stream further south and creating these cold winter weather anomalies.

Some meteorologists have pointed out that the instability of the vortex might be explained by global warming. Normally, the polar vortex is relatively stable and does not wander much. This is because the temperature difference between the polar vortex and lower latitudes is relatively large, which increases the strength of the polar vortex and causes it to be more stable. But if the temperature of the polar regions increases faster than more temperate climates, the temperature difference is reduced, lowering the strength of the vortex. This in turn pushes the jet stream further south, creating abnormal weather patterns down to Texas and Mexico.

So global warming, ironically, may be responsible for some of the freezing weather in the South.

What to Do?

So what do we do about it?

One can hope that renewable energy and conservation measures will gradually wean civilization off its dependency on fossil fuels. Perhaps a super battery would help usher in a Solar Age with fuel-efficient electric cars. Perhaps nations will become serious about confronting this problem. And perhaps by mid-century, fusion power will be online.

But if all else fails, one fallback plan is to try geoengineering to solve the problem. These are solutions to be used in a worst-case scenario.

1. Carbon sequestration

The most conservative approach is carbon sequestration, or separating out the CO_2 at the oil refinery and then burying it in the ground. On a small scale, this has already been attempted. Another idea is to separate out the CO_2 and dispose of it by mixing it with basalt found in volcanic rock. The idea is a serious one, but the bottom line is economics. Carbon sequestration costs money, and a company has to justify such an undertaking. So many companies are taking a wait-and-see position on carbon sequestration. The jury is still out on whether this will work and whether it will ever be economically viable.

2. Weather modification

When Mount St. Helens blew up in 1980, scientists were able to calculate how much volcanic ash was lofted into the environment and what the subsequent effect on the temperature was. The darkening of the atmosphere by the eruption apparently reflected more sunlight back into space, causing a cooling effect.

One might calculate how much particulate matter might be needed for a global reduction in temperature.

There are, however, dangers associated with this. Given the scale of this operation, it would be very difficult to run tests of this idea. And even if a volcanic eruption lowers the temperature temporarily by a few degrees, this is too small to avert a full climate catastrophe.

3. Algae Blooms

Another possibility is to seed the oceans, which can absorb CO_2. Algae, for example, can thrive on iron. And algae in turn absorbs CO_2. So by seeding the oceans with iron, one might be able to use algae to rein in CO_2. The problem here is that we are playing with life-forms that we do not control. Algae is not static, but can reproduce in unforeseen ways. And you cannot recall a life-form the way you would a faulty car.

4. Rain Clouds

Others have suggested modifying the weather using an old technique: silver iodide crystals. While ancient peoples might have tried to bring on the rain via dancing and incantations, nations and militaries have tried by ejecting chemicals into the atmosphere. Silver iodide crystals, for example, can hasten the condensation of water vapor, perhaps inducing rain clouds to create thunderstorms. It was believed that this method was investigated by the CIA during the Vietnam War as a way to foil enemy troops during the monsoon season by flooding them out of their sanctuaries.

Another variation is called cloud brightening, or seeding clouds so that they reflect more of the sun's energy back into space.

Unfortunately, weather modification is very local, influencing only a tiny area, while the earth's surface is very large. And the track record of seeding rain clouds is not a good one. It is very unpredictable.

5. Plant Trees

It might be possible to genetically alter plants so that they absorb more CO_2 than normal. This is perhaps the safest and most reasonable approach, but it is doubtful that enough CO_2 can be removed to reverse global warming for the entire planet. And because much of the forest land on the planet is controlled by a patchwork of nations, each with its own agenda, it would take the political will of many nations working together to embark on a plan this ambitious.

6. Calculating Virtual Weather

Given the enormous stakes involved, it is hoped that quantum computers will be able to calculate the best option. The most important task is to compile all the data to make predictions as accurate as possible.

Quantum Computers and Weather Simulation

All computer models of the weather begin by breaking up the surface of the earth into small squares or grid cells. Back in the 1990s, computer models started with square grid sizes of about 311 miles on each side. With increasing computer power, this size gets smaller all the time. (For the Fourth Assessment Report of the IPCC, Intergovernmental Panel on Climate Change, of 2007, the grid size was sixty-eight miles.)

Next, these square grids are extended into the third dimension, so they become square slabs describing various layers of the atmosphere. Typically, the atmosphere is divided into ten vertical slabs.

Once the entire earth surface and atmosphere is divided into these discrete slabs, the computer then analyzes the parameters within each slab (humidity, sunlight, temperature, atmospheric pressure, etc.). Using known thermodynamic equations for the atmosphere and energy, they then calculate how temperature and humidity vary across neighboring cells, until the entire earth is covered.

In this way, scientists can give a rough estimate of future weather. To check these results, they can be "tested" by what is called hindcasting. The computer program can be run backward in time, so that, starting with the current behavior of the weather, we can see if it can "predict" the weather in the past, when weather conditions were known with accuracy.

Hindcasting has shown that these computer models, although not perfect, have correctly "predicted" the overall weather pattern for the past fifty years. But the data is voluminous, pushing the limit of what ordinary computers can perform. Since digital computers will eventually become overwhelmed by the increasing complexity of this task, what is needed is a transition to quantum computers.

Uncertainties

No matter how powerful our computer program is, there is always the problem of unknown, unexpected factors, which are difficult to model. Perhaps the most serious uncertainty is the presence of clouds, which can reflect sunlight back into outer space, thereby reducing the greenhouse effect a bit. Since up to 70 percent of the earth's surface on average is covered by clouds, this is an important factor.

The problem is that cloud formation changes minute by minute, making long-term predictions very uncertain. Clouds are immediately affected by rapid changes in temperature, humidity, air pressure, wind currents, and other factors. Meteorologists compensate for this by taking a rough estimate of what they think cloud activity will be, given data from the past.

Another source of uncertainty is the previously mentioned jet stream. When you watch the weather report, satellite photos near the Arctic show a mass of cold air wandering its way around the globe, usually confined to the north, but sometimes going as far south as Mexico. Because the precise path taken by the jet stream is difficult to predict, meteorologists take an average estimate of the temperature shifts caused by the jet stream.

The point is that there is a limit to what digital computers can do given the uncertainties. Quantum computers, however, may be able to fix the greatest sources of uncertainty. First, quantum computers can calculate what happens if we reduce the slab size to make our predictions more accurate. The weather can rapidly change over a distance of a mile, yet these slabs are many miles across, so this introduces errors. But a quantum computer will be able to accommodate a much smaller slab size.

Second, these models estimate factors like the jet stream and clouds at fixed levels. Quantum computers will have the ability to factor in variable quantities for these parameters, so that one can simply turn a knob and change them. In this way, quantum computers will be able to construct virtual weather reports with crucial variable parameters.

We see the limit of what can be done with conventional computers when we watch the predicted path of a hurricane on TV. Estimates by different computer models are put on the screen, and you can see how much they vary. Important predictions by different computer programs, such as when and where the hurricane will make landfall and how far will it penetrate inland, often differ by hundreds of miles.

But these uncertainties, which often cost millions of dollars and the lives of innocents, will be vastly reduced when we make the transition to quantum computers.

More accurate weather reports generated by quantum computers will give us better projections, which will help us prepare for possible scenarios.

But since the burning of fossil fuels is one of the major factors driving global warming, it is important to investigate alternative sources of energy. One important source of cheap energy in the future might be fusion power, that is, harnessing the power of the sun on the earth. And the key to fusion power may be quantum computers.

CHAPTER 15

THE SUN IN A BOTTLE

S ince ancient times, people worshipped the sun as the harbinger of life, hope, and prosperity. The Greeks believed that Helios the sun god proudly rode across the sky in his blazing chariot, illuminating the world and giving warmth and comfort to the mortals below.

But more recently, scientists have tried to capture the secret of the sun and bring its limitless energy down to earth. The leading candidate for this is called fusion, which some say is like putting the sun in a bottle. On paper, it seems like the ideal solution to all our energy problems. It would generate unlimited energy forever, without many of the problems associated with fossil fuels and nuclear energy. And because it is carbon neutral, it might save us from global warming.

It seems like a dream come true.

Unfortunately, physicists oversold this technology. The joke is: every twenty years, physicists claim that fusion power is just another twenty years in the future. But now, the leading industrial nations are proclaiming that fusion power is finally within our grasp, and that it will live up to its promise of delivering unlimited energy almost for free.

Today, fusion reactors are still so expensive and complex that commercialization of this technology is probably still a few decades into the future. However, with the arrival of quantum computers, many scientists hope that some of the stubborn glitches preventing the production of fusion power may be solved, paving the way to make fusion reactors a practical and economic reality. Quantum computers may turn out to be a key technology that helps usher fusion power into our homes and cities.

The hope is that fusion power will become commercialized before global warming irreversibly heats up the planet.

Why Does the Sun Shine?

People have always wondered what powers the sun. Its energy seems limitless and even divine. Some speculated that the sun must be some gigantic furnace in the sky. But a simple calculation shows that burning fuel would only last a few centuries or millennia and also that in the vacuum of space, a fire would be instantly extinguished.

So why does the sun shine?

The secret of the sun was finally unraveled by Einstein's famous equation $E = mc^2$. Physicists realized that the sun, mainly made of hydrogen, derived its enormous energy by fusing hydrogen nuclei to form helium. When the weight of the original hydrogen was compared to the weight of helium, there was a tiny missing mass. A small fraction of the original mass was lost in the fusion process. This mass deficit, in Einstein's formula, becomes the massive energy that lights up the solar system.

The public became aware of the enormous power locked within the hydrogen atom when it was unleashed via the detonation of the hydrogen bomb. A piece of the sun, in some sense, was brought to earth, with momentous implications.

Advantages of Fusion

There are actually two ways to unleash this nuclear fire. One can fuse hydrogen together to form helium via fusion, or one can split apart the uranium or plutonium atom to release nuclear energy via fission. In each process, when you compare the weight of the ingredients to the weight of the end product, a tiny amount of mass has disappeared, which can be found in the form of nuclear energy.

Although all commercial nuclear power plants derive their energy via uranium fission, fusion has some remarkable advantages.

First, unlike fission plants, fusion does not create copious amounts of deadly nuclear waste. In a fission reactor, the uranium nucleus splits apart, releasing energy, but it also can create a cascade of hundreds of radioactive fission products, like strontium-90, iodine-131, cesium-137, and others. Some of these radioactive by-products will be radioactive for millions of years, requiring gigantic nuclear dumps to be guarded far into the future. A single commercial fission plant, for example, can create thirty tons of high-level nuclear waste in just one year. Nuclear waste dumps are like gigantic mausoleums. Worldwide, there are 370,000 tons of deadly fission products that have to be carefully monitored.

Fusion plants, by contrast, make helium gas as a waste product, which is actually commercially valuable. Some of the irradiated steel of a fusion plant might also become radioactive after decades of use, but this is easily disposed of and buried.

Second, unlike fission plants, fusion plants cannot suffer from meltdowns. In a fission plant, waste products continue to generate large quantities of heat even if a reactor is shut off. When the cooling water is lost in a nuclear fission plant accident, the temperature can soar until the reactor hits 5,000 degrees F and starts to melt, creating disastrous explosions. At Chernobyl in 1986, for example, steam and hydrogen gas explosions blew the roof off the reactor, releasing about 25 percent of the radioactive materials in the core into the atmosphere and over Europe. It was the worst commercial nuclear accident in history.

By contrast, if a fusion reactor has an accident, the fusion process simply stops. No more heat is generated, and the accident is over.

Third, the fuel for a fusion reactor is limitless. Uranium, by contrast, is in limited supply, and requires an entire fuel cycle of mining, milling, and enrichment to produce usable uranium fuel. Hydrogen, on the other hand, can be extracted from ordinary seawater.

Fourth, fusion is very efficient in releasing the energy of the atom. One gram of heavy hydrogen can produce 90,000 kilowatts of electrical energy, or the equivalent of eleven tons of coal.

And lastly, fusion and fission plants create no carbon dioxide, and hence do not exacerbate global warming.

Building a Fusion Reactor

There are two basic ingredients for a fusion machine. First, you need a source of hydrogen heated to many millions of degrees, actually hotter than the sun, turning it into plasma, which is the fourth state of matter (after solids, liquids, and gases). A plasma is a gas so hot that some of its electrons have been ripped off. It is the most common form of matter in the universe, making up stars, interstellar gas, and even lightning bolts.

Second, you need a way to contain the plasma as it is heated. In stars, gravity compresses the gas. But on earth, gravity is too weak to do this, so we use electric and magnetic fields.

The most popular design for the fusion reactor is called the tokamak, a Russian design. Start with a cylinder and then wind wire coils completely around it. Take the two ends of the cylinder and connect them together, forming a doughnut. Inject hydrogen gas into the doughnut and then shoot an electric current through the cylinder, which heats up the gas to enormous temperatures. To contain this hot plasma, huge amounts of electrical energy are fed into the coils that surround the doughnut, thereby containing the plasma with a powerful magnetic field and preventing the plasma from hitting the walls of the reactor.

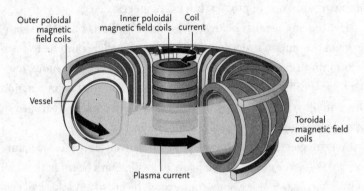

Figure 11: Tokamak
In a fusion reactor, coils of wire are wound around a doughnut-shaped chamber, creating a powerful magnetic field that confines a superhot plasma. The key to the tokamak is heating the gas so that fusion releases vast amounts of energy. In the future, quantum computers may be used to alter and even improve the precise configuration of the magnetic field, thereby increasing their power and efficiency and vastly reducing costs.

Finally, once fusion starts, hydrogen nuclei combine to form helium, releasing vast amounts of energy. In one design, two isotopes of hydrogen, deuterium and tritium, are fused together, creating energy, helium, and a neutron. This neutron, in turn, carries the energy of fusion outside the reactor, where it hits a blanket of material surrounding the tokamak.

This blanket, usually made of beryllium, copper, and steel, heats up, so that water in pipes in the blanket starts to boil. The steam created in this way can push against the blades of a turbine, causing giant magnets to spin. This magnetic field, in turn, pushes against the electrons in the turbine, generating the electricity that eventually winds up in your living room.

Why the Delays?

With all these advantages lying in wait, what is causing so many delays with fusion power? It has been about seventy years since the first fusion plants were constructed, so why is it taking so long? The problem is not one of physics, but of engineering.

Hydrogen gas must be heated to many millions of degrees, hotter than the sun, in order to make hydrogen nuclei combine to form helium and release energy. But heating gas to that enormous temperature is difficult. The gas often is unstable and the fusion reaction shuts off. Physicists have spent decades trying to contain hydrogen so that you can heat it up to stellar temperatures.

In retrospect, physicists can see how relatively easy it is for nature to release fusion power in the heart of a star. Stars begin with a ball of hydrogen gas that is evenly compressed by gravity. As this ball gets smaller and smaller, temperatures begin to rise, until it hits many millions of degrees and the hydrogen begins to fuse and the star ignites.

Notice that this process occurs naturally by itself, because gravity is monopolar, i.e., you start with one pole (not two), so the original ball of gas collapses all by itself under its own gravity. As a result, stars are relatively easy to form, and that's why we see billions of them with our telescopes.

However, electricity and magnetism are different. They are bipolar. A bar magnet, for example, always has a north and south pole. You cannot use a hammer to isolate a single north pole. If you break a magnet in half, you wind up with two smaller bar magnets, each with their own north and south poles.

So here is the problem. It is exceedingly difficult to create a powerful magnetic field to squeeze superhot hydrogen gas in the shape of a doughnut long enough to create fusion. To see why this is so difficult, think of taking a long balloon, like those used to make balloon animals. Now join the ends of the balloon so that it forms a doughnut. Then try to squeeze it evenly. No matter where you squeeze the balloon, air will manage to push out somewhere else along the balloon. It is exceedingly difficult to squeeze it so that the air inside compresses evenly.

ITER

With the ending of the Cold War and the realization that building a fusion reactor was prohibitively expensive, the nations of the world began to pool their knowledge and resources for the peaceful harnessing of the atom. In 1979, momentum began to build in the halls of the great powers for an international fusion reactor. Presidents Ronald Reagan and Mikhail Gorbachev met and helped seal the deal.

The ITER (International Thermonuclear Experimental Reactor) is an example of this international cooperation. There are thirty-five nations involved with funding this ambitious project, including the European Union, the U.S., Japan, and Korea.

To measure the efficiency of a fusion reactor, physicists introduced the quantity called Q, which is the energy generated by the reactor divided by the energy it consumes. When $Q = 1$, we hit breakeven, so it produces as much energy as it consumes. At present, the world's record for a fusion plant hovers around $Q = .7$. The ITER is projected to hit breakeven by 2025. But it is designed to eventually hit $Q = 10$, generating much more energy than it consumes.

The ITER is a monstrous machine, weighing over 5,000 tons, making it one of the most sophisticated scientific instruments of all time, alongside the International Space Station and the Large Hadron Collider. Compared to previous fusion reactor vessels, the ITER is twice as large and sixteen times as heavy. Its torus is gigantic, sixty-four feet in diameter and thirty-seven feet tall. To confine the plasma, its magnets generate a magnetic field that is 280,000 times the earth's magnetic field.

The ITER is the most ambitious fusion project in the world. It is designed to produce a net 450 million watts of energy, but won't be hooked up to the electrical grid. It will be turned on for testing in 2025 and might reach full power by 2035. If successful, it will pave the way for the next-generation fusion reactor, called DEMO, which is planned to be completed by the year 2050. DEMO is designed to hit Q = 25 and produce up to 2 gigawatts of energy.

So the goal is to have commercial fusion power before mid-century. But analysts stress that fusion power won't solve the global warming crisis any time soon. "Fusion is not a solution to get us to 2050 net zero. This is a solution to power society in the second half of this century," says Jon Amos, science correspondent for BBC News.

The key to the ITER is the huge magnetic fields, which are made possible by something called superconductivity, which is the point where all electrical resistance vanishes at super-low temperatures, thereby making possible the most powerful magnetic fields. Lowering the temperature to near absolute zero reduces resistance to electricity, eliminates waste heat, and increases efficiency of the magnetic field.

It was first discovered in 1911 when mercury was cooled down to 4.2 degrees K, near absolute zero. At that time, it was believed that random atomic motions would come to a near standstill at absolute zero, so electrons can finally freely travel without resistance. Thus it was considered strange that several substances could become superconducting at even higher temperatures. This was a mystery.

But it would take until 1957 when John Bardeen, Leon Cooper, and John Schrieffer finally created a quantum theory of superconductivity. They found that, under certain conditions, electrons can form what are called Cooper pairs and then coast on the surface of a superconductor without any resistance. The theory predicted that the maximum temperature for a superconductor was 40 degrees K.

Even before the magnets of the ITER are turned on, similar but smaller versions of the ITER have proven that the basic tokamak design is correct. ITER's design got a tremendous boost in 2022, when it was announced that two smaller versions of it, one based outside Oxford, England, and another in China, were able to achieve a new record.

The Oxford fusion reactor, called the JET (Joint European Torus), was able to hit Q = .33 for a full five seconds, breaking a record that this same reactor made twenty-four years ago. That is roughly equivalent to 11 megawatts of power, or the power to heat up sixty kettles' worth of water.

"The JET experiments put us a step closer to fusion power," says Joe Milnes, one of the directors of the laboratory. "We've demonstrated that we can create a mini star inside of our machine and hold it there for five seconds and get high performance, which really takes us into a new realm."

Arthur Turrell, an authority on fusion power, says, "It's a landmark because they managed to demonstrate the greatest amount of energy output from the fusion reactions of any device in history."

However, the Chinese made their announcement a few months later, announcing that they were able to sustain fusion for a full seventeen minutes by heating the plasma to 158 million degrees C. Their fusion reactor, called the EAST (Experimental Advanced Superconducting Tokamak), like its British counterpart is based on the original tokamak design, which indicates that the ITER is probably on the right track.

Competing Designs

Because the stakes are so high, and because large magnetic fields are so notoriously difficult to manipulate, many new ideas have been proposed to contain the plasma. In fact, there are about twenty-five upstarts fielding their own version of a fusion reactor.

In general, all tokamak fusion designs use superconductors, created by cooling the coils to near absolute zero, when electrical resistance almost vanishes. But in 1986, a new class of superconductors was found by trial and error, which was a sensational discovery. It could reach the superconducting phase at a balmy temperature of 77 degrees K. (This new class of superconductors, called high-temperature superconductors, was based on cooling down ceramics like yttrium barium copper oxide.) This was a stunning announcement, because it meant that a new quantum theory of superconductors was discovered, and that ceramics could become superconductors with ordinary liquid nitrogen. This was important since liquid nitrogen is about as expensive as milk, and would therefore considerably reduce the cost of supermagnets. (Dry ice, or solidified carbon dioxide, costs $1 per pound. Liquid nitrogen costs about $4 per pound. Liquid helium, which is what most superconductors use as coolant, however, costs $100 per pound.)

This may not sound like a big improvement to the average person, but to a physicist, this opens up a gold mine of opportunity. Since the most complex component of a fusion reactor is the magnets, this changes the entire economics and thus outlook of this technology.

Although the discovery of high-temperature ceramic superconductors came too late to be incorporated into the ITER, it opened up the possibility of using this technology for the next generation of fusion reactors.

One promising project using this new method is the SPARC reactor, which was announced in 2018 and has rapidly attracted the attention (and pocketbooks) of prominent billionaires like Bill Gates and Richard Branson, allowing SPARC to raise more than $250 million in a short amount of time. (But compared to the $21 billion spent so far on the ITER, this is pocket change.)

It passed a huge milestone in 2021 by successfully testing its high-temperature superconducting magnets, which can generate a magnetic field 40,000 times the earth's magnetic field.

"This magnet will change the trajectory of both fusion science and energy, and we think eventually the world's energy landscape," says Dennis Whyte of MIT. "It's a big deal. This is not hype, this is reality," says Andrew Holland, chief executive officer of the Fusion Industry Association. SPARC may hit breakeven Q = 1 at 2025, about the same time as the ITER, but at a fraction of the cost and time.

SPARC alone will not generate commercial electrical energy. But its successor, the ARC reactor, may. If successful, it should shift the center of gravity of fusion research, forcing the next generation of fusion reactors to adopt the very latest in technology, such as advances in high-temperature superconductors and perhaps quantum computers, which would be required to enhance the crucial stability of the magnetic field so it can contain the plasma.

However, the science of superconductors became quite confusing with the recent announcement that a room temperature superconductor was finally achieved. Normally, the creation of a room temperature superconductor would be heralded as the Holy Grail of low-temperature physics, the end product of decades of hard work. However, this discovery had one huge problem. Physicists finally created a room temperature superconductor, but only if you compressed it to 2.6 million times atmospheric pressure. To do even the simplest experiment with those astronomical pressures requires highly specialized machinery, which not everyone has. So physicists are taking a wait-and-see attitude, to see if the pressure can be reduced so that room temperature superconductors can become a useful alternative.

Laser Fusion

An entirely different approach to fusion has been taken by the U.S. Department of Energy, using gigantic laser beams instead of powerful magnets to heat up hydrogen. For a TV program I once hosted for BBC-TV, I visited the NIF (National Ignition Facility), a huge installation at the Livermore National Laboratory in California, costing $3.5 billion.

Because it is a military installation where nuclear warheads are designed, I had to go through several security checks to tour the facility. Eventually, I went past the armed guards and was ushered into the NIF control room. Even if you've seen the blueprints of the NIF on paper, you are still overwhelmed when you actually see the sheer size of this machine in person. It is truly gigantic, the size of three football fields, standing ten stories tall, dwarfing the average person.

From a distance, I could see the path taken by 192 high-powered laser beams, among the most powerful on the planet. When these laser beams are fired for a billionth of a second, they hit 192 mirrors. Each one is carefully positioned to reflect the beam onto the target, which is a small pellet the size of a pea containing lithium deuteride, which is rich in hydrogen.

This causes the surface of the pellet to vaporize and collapse, which raises its temperature to tens of millions of degrees. When heated and compressed to such a degree, fusion takes place, and tell-tale neutrons are emitted.

The eventual goal is to generate commercial energy via laser fusion. When the target is vaporized, neutrons will be emitted, which will then be sent through a blanket. As in the tokamak, the hope is that these high-energy neutrons will transfer their energy to the blanket, which then heats up and boils water, which is then fed into a turbine to generate commercial energy.

In 2021, NIF hit a milestone. It was able to produce 10 quadrillion watts of power in 100 trillionths of second, at 100 million degrees K, breaking its previous record. It compressed the fuel pellet to 350 billion times atmospheric pressure.

And finally in December 2022, NIF made headlines around the world with the sensational announcement that it had, for the first time in history, attained Q greater than 1, i.e., it generated more energy than it consumed. This was indeed a historic event, signaling the fact that fusion was a goal that can be achieved. But physicists also cautioned that this was just the first step. The second step would be to scale up the reactor so that it can power an entire city. Then after that, it must be profitably reproduced and spread around the world. It remains to be seen if NIF can be commercialized to create practical amounts of power. In the meantime, the tokamak design is still the most advanced and the most common.

Problems with Fusion

Although fusion power has the ability to change the way we consume energy on the earth, stubborn problems have led to false hopes and shattered dreams.

Many past efforts at harnessing fusion power have been disappointing. Since the 1950s, there have been over 100 fusion reactors, but none of them produced more energy than they consumed. Many were later abandoned. One fundamental problem is the toroidal (doughnut-shaped) configuration of the tokamak design. It solved one problem (the ability to contain the plasma at high temperatures) but led to another (instability).

Because of the toroidal nature of the magnetic field, it is difficult to sustain a stable fusion process for long enough to satisfy the Lawson criterion, which requires a certain temperature, density, and duration in order to create the fusion reaction.

If there are tiny irregularities in the magnetic field of the tokamak, the plasma might become unstable.

The problem is made worse by the interaction between the plasma and the magnetic field. Even if the external magnetic field can initially contain the plasma, the plasma itself has its own magnetic field, which can interact with the larger magnetic field of the reactor and become unstable.

The fact that the equations for the plasma and the magnetic field are tightly coupled creates ripple effects. If there is a slight irregularity in the magnetic field lines inside the doughnut, that, in turn, may cause irregularities in the plasma inside the doughnut. But because the plasma has its own magnetic field, it can strengthen the original irregularity. Thus, there can be a runaway effect, with the irregularity getting larger and larger each time the two magnetic fields reinforce each other. These irregularities sometimes get so large that they might touch the walls of the reactor and actually burn a hole in it. So this is the fundamental reason why it has been so difficult to satisfy the Lawson criterion and keep the fusion process stable long enough to create a self-sustaining reactor.

Quantum Fusion

This is where quantum computers come in. The equations for the magnetic field and also the plasma are both known. The problem is that these two equations are coupled to each other, so they interact with each other in complex ways. Unpredictable small oscillations can suddenly be magnified. But while digital computers have a hard time computing in this situation, quantum computers might be able to calculate with this complex arrangement.

Today, if a fusion reactor has the wrong design, it is prohibitively difficult to start all over again and redesign the reactor from scratch. However, if all the equations are inside a quantum computer, it becomes a simple matter to use the quantum computer to calculate whether the design is optimal or whether there might be more stable or efficient designs.

Changing the parameters in a quantum computer program is dramatically cheaper than redesigning an entirely new billion-dollar fusion reactor magnet.

Since a reactor can cost between $10 and $20 billion, this could result in astronomical savings. New designs can be created and tested virtually because the quantum computer can calculate its properties. In addition, a quantum computer could easily play with a series of new virtual designs to see if they improve the performance of the reactor.

The power of quantum computers can also be magnified if coupled with artificial intelligence. AI systems can vary the strength of the various magnets of a fusion reactor. Then quantum computers can analyze the flood of data from this procedure in order to increase the Q factor. For example, the AI program DeepMind has already been used to modify the fusion reactor operated by the Swiss Federal Institute of Technology in Lausanne, Switzerland.

"I think AI will play a very big role in the future control of tokamaks and in fusion science in general," says Federico Felici of the Swiss Institute. "There's a huge potential to unleash AI to get better control and to figure out how to operate such devices in a more effective way," he adds.

So AI and quantum computers can work hand in hand to increase the efficiency of fusion reactors, which in turn may energize the future and help reduce global warming.

Another application of quantum computers is to decipher how high-temperature ceramic superconductors work. As mentioned, at present no one knows how they possess this magical property. These high-temperature ceramics have been around for over forty years, yet there is no consensus. Theoretical models have been proposed, but they are just that: theoretical.

A quantum computer, however, could change this. Because the quantum computer is itself quantum mechanical, it might be able to calculate the distribution of electrons within the two-dimensional layers inside the ceramic superconductor, and hence determine which theory is correct.

In addition, we have seen that the creation of superconductors is still done by trial and error. By accident, new superconductors might be discovered. But this means that entirely new experiments must be created each time a new material is tested. There is no systematic way to find new superconductors. However, a quantum computer will be able to create a virtual laboratory in which to test new proposals for a superconductor. One might be able to rapidly test scores of interesting substances in a single afternoon, rather than taking years and spending millions to examine each one.

So quantum computers may hold the key to a pollution-free, cheap, and reliable energy future.

But if we can solve the fusion equations in a quantum computer, perhaps we can also solve the fusion equation that lies at the heart of stars, so that we can unravel the secret of the internal nuclear furnaces scattered across the night sky, how these stars explode in a supernova, and how they eventually become the most mysterious object in the universe, a black hole.

CHAPTER 16

SIMULATING THE UNIVERSE

In 1609, Galileo Galilei gazed through the telescope that he personally handcrafted and saw wonders that no one had ever seen before. For the first time in history, the true glory and majesty of the universe were unveiled.

Galileo was mesmerized by what he saw. With his own eyes, he was dazzled by a new, stunning picture of the universe as it was revealed to him each night. He was the first to see that the moon had deep craters, that the sun had tiny black spots, that Saturn had some kind of "ears" (now known as rings), that Jupiter had four moons of its own, and that Venus had phases like the moon, which proved to him that the earth revolved around the sun and not vice versa.

Galileo even organized evening skywatching parties, where the elite of Venice could see with their own eyes the true splendor of the universe. But this glorious picture did not match the one told by the religious establishment, so there was a heavy price to pay for this cosmic revelation. The Church taught that the heavens consisted of perfect, eternal celestial spheres, a testament to the glory of God, while the earth was afflicted with carnal sin and temptation. Yet Galileo could see with his own eyes that the universe was rich, varied, dynamic, and ever changing.

In fact, some historians believe that the telescope ranks as perhaps the most seditious instrument ever introduced in the history of science because it challenged the powers that be and forever altered our relationship with the world around us.

Galileo, with his telescope, was overturning everything known about the sun, moon, and the planets. Eventually, Galileo was arrested, brought to trial, and pointedly reminded that the former monk Giordano Bruno just thirty-three years earlier was burned alive in the streets of Rome for claiming that there could be other solar systems in space, some perhaps with life on them.

The revolution that was ignited by Galileo's telescope has forever altered the way we see the glories of the universe. Astronomers are no longer burned at the stake. Instead, they launch giant satellites like the Hubble and the Webb Space Telescopes to unravel the mysteries of the universe. (And there is even a statue of Bruno in the Plain of Flowers in Rome, at the very spot where he was burned alive. Every day, Bruno has his revenge as new planets are found circling distant stars in the heavens.)

Today, satellites orbiting the earth have an unparalleled view of the heavens. These instruments, like the Webb Space Telescope, which is perched up to a million miles from earth, have opened up new horizons for astronomy from their cosmic vantage point.

Science has been so successful that scientists are now drowning in an ocean of data, and quantum computers may be necessary to organize and analyze this deluge of information. Astronomers no longer shiver alone in the freezing weather, staring every lonely night through their cold telescopes, tediously chronicling the motions of each planet. Now, they program giant robotic telescopes that automatically sweep across the night sky.

Children often ask the simple question: How many stars are there? This is a difficult question to answer, but our own Milky Way galaxy has on the order of 100 billion stars. But the Hubble Telescope can, in principle, detect 100 billion galaxies. So it is estimated that there are around 100 billion times 100 billion = 10^{22} stars in the known universe.

This in turn means that an encyclopedia of all the planets that catalogued their location, size, temperature, etc., would exhaust the memory of a supercomputer. So quantum computers may be required to take the true measure of the universe.

Quantum computers may be able to sift through this astronomical tower of data to select crucial characteristics about the celestial objects. They will be able to home in on key pieces of data and extract vital conclusions from this chaotic mass with the push of a button.

Also, by calculating fusion deep inside a star, quantum computers may be able to predict when the next gigantic solar flare may paralyze the electrical grid. Quantum computers may also be able to solve the equations that can describe renegade asteroids, exploding stars, the expanding universe, and what is inside a black hole.

Killer Asteroids

There is a practical reason for analyzing these celestial bodies that hits much closer to home. Some of them may actually be dangerous, capable of destroying the earth as we know it. Sixty-six million years ago, an object about six miles across slammed into the Yucatán Peninsula of Mexico. The explosion released so much energy that it created a crater almost 200 miles in diameter, generating a tidal wave that was almost a mile high and flooded the Gulf of Mexico. It also unleashed a storm of blazing meteors, which then ignited raging infernos throughout the area. As thick dust clouds cut off sunlight and shrouded the earth in darkness, temperatures plummeted until lumbering dinosaurs could no longer hunt or eat. Perhaps 75 percent of all life-forms perished with this asteroid strike.

The dinosaurs, unfortunately, did not have a space program, so they are not here to discuss this question. But we do, and one day we may need it if an extraterrestrial object is on a collision course with earth.

So far, about 27,000 asteroids have been carefully plotted by the government and military. They are near-Earth objects (NEOs), which cross the path of the earth and hence pose a long-term threat to the planet. Most of them range in size from a football field to several miles across. But more worrisome are the tens of millions of asteroids that are smaller than a football field and are not tracked at all. They might fly undetected and cause considerable damage if they hit the earth. Another danger is long-period comets, whose location beyond Pluto is unknown, and one day they might approach the earth unannounced and undetected. So unfortunately, only a fraction of potentially dangerous objects are actually tracked by researchers.

I once interviewed the astronomer Carl Sagan, famous for his TV programs popularizing science. I asked him about the future of humanity. He replied that the earth sits in the middle of a "cosmic shooting gallery," so it was only a matter of time before one day we are confronted with a giant asteroid that might destroy the earth. That's why, he told me, we have to become a "two-planet species." That is our destiny. We should explore outer space, he said, not just to discover new worlds, but to find another safe haven in the heavens.

One asteroid that is being examined carefully as a threat is Apophis, which is roughly 1,000 feet across and will skim the earth's atmosphere in April 2029.

It will come within 10 percent of the distance between the earth and moon.

In fact, it will come so close to the earth that it will be visible to the naked eye, passing just beneath some of our satellites.

Because it will graze our atmosphere, it will encounter unpredictable atmospheric conditions, so it is impossible to say for sure what its trajectory will look like later in 2036, when it makes its next pass around the earth. It will most likely miss the earth in 2036, but that is only a guess.

The point here is that quantum computers may be necessary to track and make better approximations of the trajectory of potentially dangerous asteroids. One day, an asteroid will pass near the earth, creating mass panic as scientists try to determine if it will hit the earth or pass harmlessly by. This is where quantum computers can make a difference.

In a worst-case scenario, a distant comet from deep space may begin a long journey to our inner solar system. Without a tail, it will be invisible to our telescopes. As it whips behind our sun, sunlight finally heats up the comet's ice and a tail forms. As it suddenly emerges from behind the sun, our telescopes will at last detect the comet's tail and give us warning before a catastrophic impact. But how much warning could our telescopes give us? Perhaps a few weeks.

Unfortunately, we cannot expect Bruce Willis to come to our rescue in the Space Shuttle. First of all, the old Space Shuttle program was canceled, and the replacement for the shuttle cannot reach into deep space. But even if it could, we still wouldn't be able to intercept an asteroid and deflect it or destroy it in time.

In 2021, NASA sent the DART (Double Asteroid Redirection Test) probe into outer space to actually intercept an asteroid. For the first time in history, a human-made object succeeded in physically altering the trajectory of an asteroid. This impact will hopefully answer many questions. Is the asteroid a loose collection of rocks, which easily flies apart? Or is it a tough, solid mass that will remain intact? If it is successful, other DART-like missions will impact distant asteroids, as a dress rehearsal for what might happen one day.

In the end, it will probably be up to quantum computers to detect dangerous, planet-killing asteroids and plot their precise trajectory because there are potentially millions of them that can inflict major damage on the earth, many of them undetected.

We also need quantum computers to model the impact itself, so that we can get an estimate of how dangerous these objects might be if they actually hit the earth. An asteroid might be expected to hit the earth at velocities approaching 160,000 miles per hour, and very little is known about calculating the devastation they can unleash at these hypersonic velocities. Quantum computers may help fill the gap so we know what to expect if the earth winds up in the cross-hairs of a killer asteroid we are unable to deflect or destroy.

Exoplanets

Looking beyond our solar system, there is another reason to use quantum computers, and this is to catalogue all the planets circling other stars. Already, the Kepler space telescope and other satellites and ground-based telescopes have detected about 5,000 exoplanets in our own backyard of the Milky Way galaxy. This means that, on average, every star we see at night has a planet going around it. Perhaps roughly 20 percent of all exoplanets are earthlike, so that our own galaxy may have billions of earthlike planets beyond those we've already identified.

When I was in elementary school, I vividly remember that one of my first science books was about our solar system. After a wondrous tour around Mars, Saturn, Pluto, and beyond, the book said that there are probably other solar systems in the galaxy, and that ours is probably an average one. Probably all solar systems have rocky planets near the sun, and gas giants further out, like Jupiter, all orbiting their sun in a circular path.

Now we realize how wrong all those assumptions were. We now know that solar systems come in all sizes and shapes. Our solar system, in fact, is the oddball. We find solar systems with planets in highly elliptical orbits. We find gas giants larger than Jupiter circling extremely close to their sun. We find solar systems with multiple suns.

So one day, when we have an encyclopedia of the planets in the galaxy, we will be amazed at their rich variety. If you can imagine a strange planet, there is probably one like it out there.

We will need a quantum computer to keep track of all the possible paths that describe planetary evolution. As we launch more telescopes into space, this encyclopedia of planets will explode in size, requiring immense computing power to analyze their atmospheres, chemical composition, temperature, geology, wind patterns, and other characteristics that will generate mountains of data.

ET in Space?

One goal that quantum computers will focus on is the search for other intelligent life-forms. An embarrassing question arises: How will we recognize intelligence that might be totally alien to ours? Will we recognize an alien life-form if it is right in front of us? We may need quantum computers to recognize patterns that might be totally hidden from conventional computers.

Astronomer Frank Drake, in the 1950s, came up with an equation that tried to estimate how many advanced civilizations there might be in the galaxy. You start with the 100 billion stars in the galaxy, and reduce that number with a series of reasonable assumptions. You reduce it by the fraction that have planets, the fraction that have planets with atmospheres, the fraction that have planets with atmospheres and oceans, the fraction that have planets with microbial life, etc. No matter how many or how few reasonable assumptions you make about these planets, the final number is usually in the thousands.

Yet the SETI (Search for Extraterrestrial Intelligence) project has found no evidence of any intelligent radio signal from outer space. None whatsoever. Their powerful radio telescopes in Hat Creek outside San Francisco only record dead silence or static. So we are left with the Fermi paradox: If the probability of intelligent alien life in the universe is so high, then where are they?

Quantum computers may help answer that question. Since they excel at poring over huge amounts of data to find hidden clues, and artificial intelligence excels at learning to identify new things by picking up patterns, combined they might learn to plow through massive amounts of data to find what may be hiding there, even if it is bizarre or totally unexpected.

I got a taste of this when I hosted a program for the Science Channel about alien intelligence, where we analyzed the intelligence of nonhumans, like the dolphin. I was placed in a swimming pool with several playful ones. The goal was to have them communicate with each other to see if we could measure their intelligence. In the water were sensors that could record all their chirps and squeals.

How can a computer find signs of intelligence in this mass of apparent noise and gibberish? Tape recordings like these can be run through a computer program designed to look for specific patterns. For example, the letter of the alphabet which is the most commonly used in English is "e." Examining someone's writing, it is possible to rank each letter by how often it is used. This ranking of the letters of the alphabet, on the basis of how often you use them, is specific to you. Two different people will use a slightly different ranking of letters. This can, in fact, be used to find forgeries. For example, by running the works of Shakespeare through this program, one can tell if any of his plays were written by someone else.

When the recordings of the dolphins were analyzed by the computer, at first you only hear a random jumble of sounds. But it was specifically designed to find out how often certain sounds were heard. The computer eventually concluded that there was a rhyme and reason behind all the chirps and squeals.

Other animals have been tested in the same way, and there is a drop-off in intelligence as we go to more primitive organisms. In fact, by the time one analyzes insects, these signs of intelligence drop to near zero.

Quantum computers can sift through this vast collection of data to find interesting signals, and AI systems can be trained to look for unexpected patterns. In other words, AI and quantum computers, working together, may be able to find evidence of intelligence even in a jumble of chaotic signals from space.

Stellar Evolution

Another immediate application of quantum computers is to fill the gaps in our understanding of stellar evolution and the life cycle of stars, from their birth to ultimate death.

When I was getting my PhD in theoretical physics from the University of California at Berkeley, my roommate was getting his PhD in astronomy. Every day, he would say goodbye and declare that he was going to bake a star in an oven. I thought he was joking. You can't bake stars. Many are bigger than our sun. So one day I finally asked him what he meant by baking a star. He thought for a moment and then told me that the equations describing stellar evolution are not complete, but are good enough so that they can simulate the life cycle of a star from birth to death.

In the morning, he would input the parameters of a dust cloud of hydrogen gas into the computer (such as the size, the gas content, the temperature of the gas). Then the computer would calculate how the gas cloud would evolve. By lunchtime, the cloud would collapse under gravity, heat up, and ignite into a star. By the afternoon, it would blaze away for billions of years and act like a cosmic oven, fusing or "cooking" hydrogen and then creating increasingly heavy elements, such as helium, lithium, and boron.

We have learned a lot from simulations like these. In the case of our sun, after 5 billion years it will have exhausted most of its hydrogen fuel and will start to burn helium. At that point, it will start to expand enormously, becoming a red giant so large that it will fill the sky and stretch across the entire horizon. It will engulf planets out to Mars. The sky will be on fire. The oceans will boil, the mountains will melt, and all will go back into the sun. From star dust we came, and to star dust we will return.

As the poet Robert Frost once wrote,

> Some say the world will end in fire
> Some say in ice.
> From what I've tasted of desire
> I hold with those who favor fire.
> But if it had to perish twice
> I think I know enough of hate
> To say that for destruction ice
> Is also great
> And would suffice.

Eventually, the sun will exhaust its helium and will shrink into a white dwarf star, which is only the size of the earth but weighs almost as much as the original sun. As it cools, it will become a dead, black dwarf star. So that is the future of our sun, to die in ice, rather than fire.

However, for truly massive stars in the red giant phase, they will continue to fuse higher and higher elements, until eventually they hit the element iron, which has so many protons that they repel each other and hence fusion finally stops. Then without fusion, the star collapses under gravity, and temperatures can soar to trillions of degrees. At that point, the star explodes into a supernova, one of the greatest cataclysms in nature.

So a giant star can die in fire, not ice.

Unfortunately, there are still many gaps in calculating the life cycle of stars, from gas clouds to a supernova. But with quantum computers modeling the fusion process, perhaps many of them can be filled.

This may be a crucial piece of evidence as we face another ominous threat: a monster solar flare that could throw civilization hundreds of years into the past. To predict the occurrence of a deadly solar flare, you need to know the dynamics deep inside a star, which is far beyond the capability of a conventional computer.

Carrington Event

For example, we know very little about the interior of our sun, and hence are vulnerable to catastrophic bursts of solar energy that send huge amounts of superhot plasma into outer space. We were reminded of how little we know about the sun in February 2022, when a giant burst of solar radiation hit the earth's atmosphere and wiped out forty of forty-nine communications satellites sent into orbit by Elon Musk's SpaceX. This was the largest solar disaster in modern history, and it is likely to happen again, since we have much to learn about these corona mass discharges.

The biggest solar flare in recorded history, called the Carrington Event, took place in 1859. Back then, this monster solar flare caused telegraph wires to catch on fire over much of Europe and North America. It created atmospheric disturbances all over the planet, with the aurora borealis blanketing the night sky over Cuba, Mexico, Hawaii, Japan, and China. You could read the newspaper at night in the Caribbean by the light of the aurora. In Baltimore the aurora was brighter than a full moon. One gold miner, C. F. Herbert, wrote this graphic eyewitness account of this historic event:

> A scene of almost unspeakable beauty presented itself.... Lights of every imaginable color were issuing from the southern heavens, one color fading away only to give place to another if possible more beautiful than the last.... It was a sight never to be forgotten, and was considered at the time to be the greatest aurora recorded.... The rationalist and pantheist saw nature in her most exquisite robes.... The superstitious and the fanatical had dire forebodings, and thought it a foreshadowing of Armageddon and final dissolution.

The Carrington Event happened when the Electric Age was in its infancy. Since then, there have been attempts to reconstruct the data and then estimate what might occur if another Carrington Event were to happen in modern times. In 2013, researchers at Lloyd's of London and Atmospheric and Environmental Research (AER) in the U.S. concluded that another Carrington Event could cause up to $2.6 trillion in damages.

Modern civilization could come to a grinding halt. It would knock out our satellites and the internet, cause short circuits in power lines, paralyze all financial communications, and cause global blackouts. We would be thrown perhaps 150 years into the past. Rescue and repair teams would fail to come to the rescue, because they too would be caught in the global blackout. With perishable food rotting, eventually this could trigger massive food riots and a disintegration of social order and even governments, as people desperately scavenge for scraps of food.

Will it happen again? Yes. When might it happen? No one knows. One clue might come from analyzing previous Carrington-type events. Studies have been done looking at the concentration of carbon-14 and beryllium-10 in ice cores, hoping to find evidence of prehistoric solar flares. Studies have shown possible ones that erupted in 774–75 CE and 993–94 CE. In fact, the ice core data of the 774–75 CE event indicates that it was perhaps ten times more energetic than the Carrington Event. (And the solar eruption of 993–94 CE was so intense that it left its mark in ancient wood, which historians have used to date the early Viking settlements in the Americas.) But back then, before the dawning of the Electric Age, civilization barely took notice.

The largest solar flare in recent history took place in 2001. A huge coronal mass ejection was sent hurtling 4.5 million miles per hour into space. Fortunately, the flare missed the earth. Otherwise it could have inflicted widespread damage across the planet comparable to the Carrington Event.

Scientists have pointed out that it might be possible to prepare for the next Carrington Event if we allocate funds to reinforce our satellites, shield delicate electronics, and build redundant power stations. This would be a tiny down payment to prevent a catastrophic loss of our electrical system. But usually these warnings are ignored.

Physicists know that coronal mass discharges occur when the magnetic lines of force on the surface of sun cross each other, spewing enormous amounts of energy into space. But what happens inside the sun to create these conditions is not known. The basic equations for plasmas, thermodynamics, fusion, convection, magnetism, and so forth are known, but solving them as they occur in the interior of the sun surpasses the ability of modern computers.

So one day quantum computers may be able to unravel the complex equations inside the sun and help predict when the next giant solar flare might threaten civilization. We know that there must be huge convection currents of superhot plasma churning deep inside the sun, but we have no idea when the next solar flare might erupt or whether it will hit the earth. So if a quantum computer can "cook" stars in its memory, then we may be able to prepare for the next Carrington Event.

But quantum computers can go even further, ultimately solving the greatest cataclysm in the universe. The Carrington Event could paralyze a continent, but a gamma-ray burst could do far worse, incinerating an entire solar system.

Gamma-Ray Bursts

In 1967, a mystery unfolded in outer space. The Vela satellite, which was specifically launched by the U.S. to detect unauthorized detonations of nuclear bombs, picked up strange radiation from a huge burst of gamma rays. This gigantic blast came from an unknown source, sparking a deadly serious guessing game. Were the Russians testing an unknown weapon of unparalleled power? Was an emerging nation testing a new breakout weapon? Was this a massive failure of U.S. intelligence?

Alarm bells went off in the Pentagon. Immediately, top scientists were asked to identify this anomaly and determine where it came from. Soon afterward, other bursts of gamma rays were detected. Pentagon planners breathed a sigh of relief when their origin was finally nailed down. They were coming not from the Soviet Union but from distant galaxies. Scientists were amazed to find that these bursts only lasted a few seconds, but released more radiation than an entire galaxy. In fact, they released more energy than the sun will generate in its entire 10-billion-year history. They were the biggest explosions in the entire universe, second only to the Big Bang itself.

Since these gamma-ray bursts usually last for a few seconds before they fade away, it meant that an early warning system would be difficult to create. But eventually, a network of satellites was designed to detect these events as soon as they occur and immediately alert detectors on earth to zero in on them in real time.

There are many gaps in our understanding of gamma-ray bursts, but the leading theory is that they are either collisions between neutron stars and black holes, or stars collapsing into black holes. They may represent the final stages in the life of stars. So quantum computers might be necessary to explain precisely why so much energy is released by stars as they reach the endpoint of their life cycle.

Some of these potential dangers from exploding stars are not far from the earth. In fact, some of the atoms in your body may have been "cooked" by an ancient supernova billions of years ago. As we mentioned earlier, stars like our sun by themselves do not have enough heat to create elements beyond iron, such as zinc, copper, gold, mercury, and cobalt. These elements were created in the heat of a supernova explosion that took place billions of years before our sun was born. So the very presence of these elements in our body is evidence that a supernova took place in our neighborhood of the galaxy. In fact, some scientists have speculated that the Ordovician extinction 500 million years ago, which wiped out 85 percent of all aquatic life on earth, was triggered by a nearby gamma-ray burst.

Closer to home, the red giant star Betelgeuse, which is 500–600 light-years from earth, is unstable and will at some point undergo a supernova explosion. It is the second-brightest star in the constellation Orion. When it finally explodes, it is close enough that it will probably outshine the moon at night, and even cast a shadow. Recently, it had noticeable changes in brightness and shape, causing some speculation that it is on the verge of exploding, but this is still fiercely debated.

The point, however, is that there is much that we do not understand about supernovae, and the gaps may be filled with quantum computers. One day quantum computers will be able to explain the entire life history of stars, including the sun, and also potentially dangerous unstable stars in our vicinity.

But it is the end product of a supernova that has generated much interest—black holes.

Black Holes

Simulating black holes can quickly exhaust the calculational power of an ordinary digital supercomputer. For a large star, perhaps ten to fifty times more massive than our sun, there is the possibility that it will explode as a supernova, turn into a neutron star, and perhaps collapse into a black hole. No one really knows what happens when a massive star collapses gravitationally, because the laws of Einstein and the quantum theory begin to fail, and new physics is surely required.

For example, if we simply follow the mathematics of Einstein, the black hole will collapse behind a mysterious dark sphere, called the event horizon. This was actually photographed in 2021 by lashing together the light from a series of radio telescopes around the earth, creating a radio telescope effectively the size of the planet itself. It revealed that the event horizon at the heart of the galaxy called M87, about 53 million light-years from earth, was a dark sphere encircled by superhot luminous gases.

What lies inside the event horizon? No one knows. It was once thought that a black hole might collapse into a singularity, a super-compact point of unimaginable density. But that picture has changed, since we see black holes rotating at tremendous velocities. Instead of a pinpoint, physicists now believe that black holes may collapse into a spinning ring of neutrons instead, where the usual concepts of space and time are turned upside down. The mathematics states that if you fall through the ring, you might not die at all, but enter a parallel universe. So the spinning ring becomes a wormhole, a gateway to another universe beyond the black hole.

The spinning ring acts very much like Alice's looking-glass. On one side, you have the gentle countryside of Oxford. But if you pass through the looking-glass, you enter the parallel universe of Wonderland.

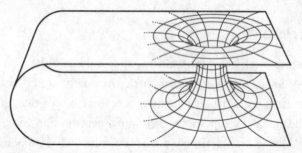

Figure 12: Quantum computers and black holes
According to relativity, a spinning black hole might collapse into a ring of neutrons, which can connect two different regions of space-time, creating a wormhole or a gateway between two universes. But a quantum computer may be necessary to determine how stable they are under quantum corrections.

Unfortunately, the mathematics of black holes cannot be trusted, because quantum effects have to be included as well. Quantum computers may be able to give us simulations of Einstein's theory and the quantum theory when space and time are twisted at the center of a black hole. In these conditions, the equations are highly coupled. First, we have the energy due to gravity and the folding of space-time. Then we have the energy due to various subatomic particles. But these particles, in turn, have their own gravitational field, which mixes with the original field in complex ways. So we have a tangle of equations, each one affecting the others, in a highly intricate mix that is beyond the reach of conventional computers, but perhaps not quantum computers.

But quantum computers may also help to answer an embarrassing, age-old question. What is the universe made of?

Dark Matter

After 2,000 years of speculation and countless experiments, we still cannot answer the simple question asked by the Greeks: What is the world made of?

Most elementary school textbooks claim that the universe is mainly made of atoms. But that statement is now known to be wrong. The universe is actually mainly made of mysterious invisible dark matter and energy. Most of the universe is dark, beyond the ability of our telescopes to study it and our senses to detect it.

Dark matter was first theorized by Lord Kelvin in 1884. He noticed that the amount of mass necessary to account for the spin of the galaxy was much larger than the actual mass of the stars. He concluded that most stars were actually dark, that they were not luminous. More recently, astronomers like Fritz Zwicky and Vera Rubin confirmed this strange observation, realizing that the galaxy and stellar clusters are spinning too fast and according to our equations should therefore fly apart. In fact, our galaxy spins about ten times faster than expected. But because of the tremendous faith that astronomers had in Newton's theory of gravity, this result was largely ignored.

Over the decades, it was found that not just the Milky Way but all galaxies exhibited this same curious phenomenon. Astronomers began to realize that the galaxies contained invisible dark matter that held them together. This halo was many times more massive than the galaxy itself. Most of the universe, it seemed, was made of this mysterious dark matter.

(Even more mysterious is dark energy, which is a strange form of energy that fills the vacuum of space and even causes the universe to expand. Although dark energy makes up 68 percent of the known matter/energy content of the universe, almost nothing is known about it.)

This chart summarizes the latest data of what scientists think the world is made of:

Dark energy	68 percent
Dark matter	27 percent
H and He	5 percent
Higher elements	.1 percent

We now realize that many of the elements that make up our body only represent about .1 percent of the universe. We are truly anomalies. But the stuff that makes up most of the universe possesses strange properties. Since dark matter does not interact with ordinary matter, if you held it in your hand it would sift right through your fingers and fall to the floor. But it wouldn't stop there; it would continue to fall through the dirt and concrete, as if the earth were not there. It would keep falling through the crust of the earth and sail on to China. There, it would gradually reverse direction by the pull of earth's gravity and travel back the way it came, until it finally reached your hand again. Then it would oscillate back and forth through the planet.

Today, we have maps of this invisible matter. The way we determine the presence of invisible dark matter is the same way that you know there is glass in your eyeglasses. Glass distorts light, so you can observe its effects. Dark matter distorts light in much the same way.

So by correcting for the refraction of light through dark matter, we can generate 3D maps of it. Sure enough, we find that dark matter concentrates around galaxies, holding them together.

But embarrassingly, we don't know what dark matter is made of. It is apparently made of a substance never seen before, something that lies outside the Standard Model of subatomic particles.

So the key to solving the mystery of dark matter may be to understand what lies beyond the Standard Model of particles.

Standard Model of Particles

Quantum computers, as we have seen, exploit the counterintuitive laws of quantum mechanics to do their calculations. But quantum mechanics itself has not been idle. It has evolved as larger particle accelerators have smashed protons into each other to find out the basic constituents of matter. At present, the most powerful accelerator in the world is the Large Hadron Collider, outside Geneva, Switzerland, the largest scientific machine ever built. It is a tube 16.6 miles around, with magnets so powerful they can hurl protons to 14 trillion electron volts.

For a BBC series I once hosted, I visited the LHC and even touched the tube at the heart of the accelerator when it was still being built. It was a breathtaking experience, knowing that, in a few more years, protons would be hurtling inside this tube with mind-blowing energies.

After decades of hard work with the LHC, physicists have finally converged on something called the Standard Model, or the Theory of Almost Everything. The old Schrödinger equation, we saw, could explain the interaction of electrons with the electromagnetic force. The Standard Model, however, could unify the electromagnetic force with the strong and weak nuclear forces as well.

So the Standard Model of particles represents the most advanced version of the quantum theory. It is the culmination of the work of scores of Nobel Prize winners and the end product of billions of dollars spent on gigantic atom smashers. By rights, it should be a shining landmark to the noblest achievement of the human spirit.

Unfortunately, it is a mess.

Instead of being the finest product of divine inspiration, it is a rather crude hodgepodge of particles. It consists of a bewildering collection of subatomic particles that do not appear to have much rhyme or reason. It has thirty-six quarks and antiquarks, over nineteen free parameters that can be adjusted at will, three generations of identical particles, and a bunch of exotic particles called gluons, W and Z bosons, Higgs bosons, and Yang-Mills particles, among others.

It is a theory only a mother can love. It's like putting an aardvark, platypus, and whale together with Scotch tape and calling it nature's finest creation, the end product of millions of years of evolution.

Worse, the theory makes no mention of gravity and cannot explain dark matter and dark energy, which make up the vast majority of the known universe.

There is only one reason why physicists study this awkward theory: it works. It undeniably describes the low-energy world of subatomic particles like mesons, neutrinos, W bosons, and so on. The Standard Model is so awkward and ugly that most physicists feel it is just the lowest-energy approximation of a more beautiful theory that exists at higher energies. (To paraphrase Einstein, if you see the tail of a lion, you suspect that sooner or later a lion will emerge.)

But for about the past fifty years, physicists have seen no deviation from the Standard Model.

Until now.

Beyond the Standard Model

The first inkling of a crack in the Standard Model came from the Fermi National Accelerator Laboratory outside Chicago in 2021. The huge particle detector there found a slight deviation in the magnetic properties of mu mesons (which are commonly found in cosmic rays). A massive amount of data had to be analyzed to find this tiny deviation, but if it holds up, it could signal the presence of new forces and interactions beyond the Standard Model.

It could mean that we are getting a glimpse of the world beyond the Standard Model, where a new physics may emerge, perhaps string theory.

Quantum computers excel as search engines, finding that elusive needle in the haystack. Many physicists believe that our particle accelerators will eventually pick up conclusive evidence of particles beyond the Standard Model, which will reveal the true simplicity and beauty of the universe.

Already, quantum computers are being used by physicists to understand the mysterious dynamics of particle interactions. At the LHC, two beams of high-energy protons are slammed into each other with an energy of 14 trillion electron volts, creating energies not found since the beginning of the universe. This titanic collision creates a gigantic shower of subatomic debris. A staggering trillion bytes per second of data is created by this colossal collision, which is then analyzed by a quantum computer.

Beyond that, physicists are already drafting plans for a replacement for the Large Hadron Collider, called the Future Circular Collider, to be built at CERN in Switzerland. At sixty-two miles in circumference, it will dwarf the 16.6-mile LHC. It will cost $23 billion and will reach the astronomical energy of 100 trillion electron volts. It will be by far the biggest scientific machine on the planet.

If it is built, it will re-create the conditions when the universe was born. It should take us as close as humanly possible to the ultimate theory, the Theory of Everything, that Einstein searched for in the last thirty years of his life. The flood of data emerging from this machine will overwhelm any conventional computer. In other words, perhaps the secret of creation itself may be unraveled by a quantum computer.

String Theory

So far, the leading (and only) candidate for a quantum theory beyond the Standard Model is string theory. All competing theories have been shown to be divergent, anomalous, inconsistent, or missing crucial aspects of nature. Any one of these defects would be fatal to a physical theory.

(I get plenty of emails from people who claim that they have finally found this Theory of Everything. I tell them that there are three criteria that your theory must obey:

1. It must contain Einstein's theory of gravity.
2. It must contain the entire Standard Model of particles, with all its quarks, gluons, neutrinos, etc.
3. It must be finite and free of anomalies.

So far, the only theory that can satisfy these three simple criteria is string theory.)

String theory says that all elementary particles are nothing but musical notes on tiny vibrating strings. Like a rubber band that can oscillate at different frequencies, string theory says that each vibration of this tiny rubber band corresponds to a particle, so the electron, quark, neutrino, and all the other players in the Standard Model are nothing but different musical notes. Physics then corresponds to the harmonies that one can play on these strings. Chemistry corresponds to the melodies created by vibrating strings. The universe can be compared to a symphony of strings. And lastly, the "mind of God" that Einstein wrote about would correspond to cosmic music resonating through the universe.

Remarkably, when calculating the nature of these vibrations, one can find gravity, which is the force conspicuously missing in the Standard Model. Thus, string theory gives us a credible reason for believing that it may be the Theory of Everything. (In fact, if Einstein had never been born, general relativity would have been discovered as a by-product of string theory, as nothing but one of the lowest notes of the vibrating string.)

But if this theory can unify both the theory of gravity as well as subatomic forces, then why have Nobel Prize winners split on this theory, with some saying it is a dead end, while others are saying that this could be the theory that eluded Einstein? One problem is its predictive power. Not only does it contain the Standard Model of particles, it includes much more. In fact, it may have an infinite number of solutions, an embarrassment of riches. If so, then which solution describes our universe?

On one hand, we realize that all great equations have an infinite number of solutions. String theory is no exception. Even Newton's theory can explain an infinite number of things, such as baseballs, rockets, skyscrapers, airplanes, etc. You have to specify ahead of time what you are investigating, i.e., you have to specify the initial conditions.

But string theory is a theory of the entire universe. Thus, you have to specify the initial conditions of the Big Bang. But no one knows the conditions that set off the initial cosmic explosion that created the universe.

This is called the landscape problem, that there seems to be an infinite number of solutions to string theory, creating a vast landscape of possibilities. Each point on this landscape corresponds to an entire universe. One of these points can explain the features of our universe.

But which one is ours? Is string theory a theory of everything, or a theory of anything?

At present, there is no consensus to resolve this problem. One solution might be to create a new generation of particle accelerators, such as the Future Circular Collider mentioned earlier, the Circular Electron Positron Collider that China has proposed, or the International Linear Collider from Japan. But there is no guarantee that even these ambitious projects will resolve this important question.

Quantum Computers May Hold the Key

My own point of view is that perhaps quantum computers may offer the ultimate answer to this question. We saw earlier how in photosynthesis, nature uses the quantum theory to survey a vast collection of paths with the principle of least action. One day, it might be possible to put string theory onto a quantum computer to select out the correct path. Perhaps many of the paths found in the landscape are unstable and quickly decay, leaving only the correct solution. Perhaps our universe emerges as the only stable one.

So quantum computers may be the final step in finding the Theory of Everything.

There is some precedent for this. The theory that best describes the theory of the strong nuclear force is called quantum chromodynamics (QCD). It is a theory of subatomic particles that binds the quarks together to create the neutron and the proton. Originally, it was thought that physicists would be clever enough to solve QCD completely using pure mathematics. This proved to be an illusion.

Today, physicists have pretty much given up trying to solve QCD by hand, and instead they rely on gigantic supercomputers to solve these equations. This is called Lattice QCD, which divides up space and time into billions of tiny cubes, forming a lattice. One solves the equations for one tiny cube, uses that to solve the equations for the next neighboring cube, and repeats the same process for all that follow. In this way, eventually the computer solves for all the neighboring cubes, one after the other.

Similarly, one may have to resort to quantum computers to eventually solve for all the equations of string theory. One hope is that the true theory of the universe may arise from this process. So quantum computers may hold the key to creation itself.

CHAPTER 17

A DAY IN THE YEAR 2050

January 2050, 6:00 a.m.

Your alarm clock is ringing, and you wake up with a splitting headache.

Molly, your personal robotic assistant, suddenly appears in your wall screen. She cheerfully announces, "It's now six a.m. Remember, you told me to wake you up."

Sleepily, you reply, "Oh, my head hurts. What was I doing last night to deserve this?"

Molly says, "Remember, you were at the party celebrating the opening of the new fusion reactor. You must have had too much to drink."

Slowly, it all comes back to you. You remember that you are an engineer for Quantum Technologies, one of the biggest quantum computer firms in the country. Quantum computers seem to be everywhere these days, and the party last night was to celebrate the opening of the latest fusion reactor, a landmark event made possible by quantum computers.

You recall that a reporter at that party asked you, "What's all the excitement about? Why the big fuss over hot gas?"

You replied, "Quantum computers finally determined how to stabilize the hot gas inside a fusion reactor, so that an almost unlimited amount of energy can be extracted from fusing hydrogen into helium. This could be the key to the energy crisis."

This means that there will be scores of fusion reactors opening up around the world, and plenty more parties to get drunk at. A new era of cheap, renewable energy is opening up, due to quantum computers.

But now it's time to catch up with the news. You tell Molly, "Please turn on the morning news concerning developments in science."

The wall screen suddenly turns on. Whenever you hear the latest news, you like to play a game with yourself. After hearing each science story, you want to identify which of those stories, if any, are *not* made possible by quantum computers.

The video host declares, "A new fleet of supersonic jets has been approved by the government, vastly reducing the time to cross the Pacific and Atlantic Oceans."

You realize that it was quantum computers that, through the use of virtual wind tunnels, found the right aerodynamic design that eliminated the noise from sonic booms, which have helped make possible this new flood of supersonic jetliners.

The host next announces, "Our astronauts on Mars have successfully built a large solar panel and a bank of super batteries to store energy for the red planet colony."

You know that all this was made possible by the quantum computers that created the super battery powering the Mars outpost. But it has also reduced our dependence on coal and oil plants on earth.

Next, the host announces, "Doctors around the world are heralding a new Alzheimer's drug, which can prevent the buildup of the amyloid protein that causes this fateful disease. This result could affect the lives of millions."

You are proud that your company was at the forefront of using quantum computers to isolate the specific type of amyloid protein responsible for Alzheimer's disease.

Hearing the science news, you smile to yourself, because, once again, all the science stories have been made possible, directly or indirectly, by quantum computers.

After listening to the news, you drag yourself to the bathroom, take a shower, and brush your teeth. As you watch the water go down the drain, you realize that your wastewater is being silently sent to a biolab, where it is being analyzed for cancer cells. Millions of people are blissfully unaware of the fact that they are having a thorough medical checkup several times a day with a quantum computer that is silently connected to their bathroom.

Because quantum computers can now identify cancer cells years before a tumor forms, cancer has been reduced to something like the common cold. Because cancer runs in your family, you think to yourself, "Thank God cancer is no longer the killer it used to be."

Finally, as you get dressed, the wall screen lights up again. This time, the image of your AI doctor lights up the wall screen.

"So what is it this time, doc? Any good news, I hope?"

Robo-doc, your personal robotic doctor, says, "Well, I have some good news and some bad news. First, the bad news. Analyzing the cells in your wastewater from last week, we have determined that you have cancer."

"Wow, so that's the bad news; then what's the good news?" you ask anxiously.

"The good news is that we've located the source and found only a few hundred cancer cells growing in your lung. Nothing to worry about. We've analyzed the genetics of the cancer cells and will give you a shot to boost your immune system to defeat this cancer. We just received the latest shipment of genetically modified immune cells that were created by quantum computers from your company to attack this particular cancer."

You are relieved. Then you ask him another question. "Be honest with me. If your quantum computers had not detected cancer cells in my bodily fluids, then what might have happened, say, ten years ago?"

Robo-doc answers, "A few decades ago, before quantum computers became widespread, you would have had several billion cancer cells growing in a tumor in your body by now, and you would have died in about five years."

You gulp. You feel proud to work for Quantum Technologies.

Molly suddenly interrupts Robo-doc. "This message just came in. There is an urgent meeting in the home office. Your presence is requested, immediately, in person."

"Uh-oh," you say to yourself. Usually, most mundane tasks are done online. But this time they want all hands on deck, in person. It must be an important meeting.

You tell Molly, "Cancel my appointments and send for my car."

Your driverless car arrives a few minutes later, which takes you to your office. The traffic is not so bad, because millions of sensors embedded in the road are connected to quantum computers, which adjust each traffic light, second by second, to eliminate bottlenecks.

When you arrive, you get out of the car and say, "Go park yourself. And be ready to pick me up afterward on a moment's notice." Your car connects to the quantum computer monitoring all traffic in the city, identifies the nearest empty parking slot.

You enter the conference room and can see the biographies of those sitting around you in your contact lens. The big shots in the company are all there. It must be an important meeting.

The president of the company is addressing this distinguished group of executives.

"I am shocked to announce that this week, our quantum computers have detected a virus never seen before. Our international network of sensors in sewer systems is our first line of defense against deadly viruses, and it detected a new virus near the border with Thailand. This virus has caught us off guard. It's highly lethal and highly contagious, probably originated in a bird of some sort. I don't have to remind you that the last pandemic cost over one million lives in the U.S. and almost tanked the world economy. I've handpicked a group of our top people to fly immediately to Asia to analyze the threat. We have our supersonic transport ready for takeoff. Any questions?"

Hands shoot up. Many of the questions are in a foreign language, but your contact lens translates them into English.

You were looking forward to a nice, quiet weekend. All your plans are now out the window. This time, a flying car takes you to the airport, where a supersonic transport is waiting for you. You have breakfast in New York, lunch over Alaska, and dinner in Tokyo, followed by an evening meeting. "Supersonic jets are a great improvement over conventional jets, with their agonizing thirteen-hour trip from New York to Tokyo," you muse to yourself.

Then you remember that back in elementary school, you read stories in history books of the nightmare caused by the Pandemic of 2020, when the world was totally unprepared to deal with an unknown virus. In fact, it killed some of your relatives. But this time, all the pieces are in place.

The next day, you are given a briefing. Your manager says, "Fortunately, quantum computers were able to identify the genetics of this virus, locate its molecular weak spots, and produce plans for vaccines which will be effective against this disease. All this was done in record time because of quantum computers, which could also analyze all airplane and train records to see how the virus may have spread internationally. Sensors at all major airports and train stations have now been calibrated to pick up the unique smell of this new virus."

After a week touring the company's laboratories, you fly back to New York, confident that your team has the new virus under control. You take pride in the fact that your efforts might have saved a few million lives and prevented the collapse of the world economy.

Back home, you ask Molly for your latest appointments. "Well, this time we have a request for an interview with you from one of the biggest magazines on the planet. They are doing a feature story on quantum computers. Shall I set it up?"

You are pleasantly surprised when the journalist arrives at your office. Sarah is prepared, knowledgeable, and very professional.

Sarah says, "I hear that quantum computers seem to be everywhere these days. Old digital computers, like dinosaurs, are being thrown into the junkyard. Everywhere I go, it seems that quantum computers are replacing the older generation of silicon computers. Every time I am on my cell phone, they tell me that I am actually talking to a quantum computer somewhere in the cloud. But tell me, with all this progress, will this help solve our pressing social problems? I mean, let's be real. For example, will this help feed the poor?"

You fire right back. "Well, actually, the answer is yes. Quantum computers have unlocked the secret of taking nitrogen in the air that we breathe every day and converting it into the ingredients for fertilizers. This is creating a Second Green Revolution. The naysayers used to claim that, with the population explosion, there will be starvation, wars, mass migrations, food riots, and so on. None of these have happened, due to quantum computers—"

"But wait a minute," interrupts Sarah. "What about all the problems with global warming. Just blink, and the internet in your contact lens has images of massive forest fires, droughts, hurricanes, flooding. . . . The weather seems to have gone wild."

"Yes," you admit, "industry spewed massive amounts of CO_2 into the atmosphere over the last century, and we are finally paying the price. All the predictions have come true. But we are fighting back. Quantum Technologies has been at the forefront of creating a super battery that can store vast amounts of electrical energy, greatly decreasing the cost of energy and helping usher in the long-awaited Solar Age. We now have power when the sun doesn't shine and the winds don't blow. The energy from renewable technologies, including from fusion plants that are now opening up around the world, is now cheaper than energy from fossil fuels, for the first time in history. We are turning the corner on global warming. Let's hope we are in time."

"Now, let me ask you a personal question. How have quantum computers affected your family and loved ones?" Sarah asks.

You reply sadly, "My family has suffered greatly from Alzheimer's disease. I saw it firsthand with my mother. At first, she would forget things that happened a few minutes ago. Then she gradually became delusional, talking about things that never happened. Next, she forgot the names of all her loved ones. Finally, she even forgot who she was. But I am proud to say, quantum computers are now solving the problem. At the molecular level, quantum computers have isolated the precise misshapen amyloid protein that gums up the brain. A cure for Alzheimer's is at hand."

Next she asks, "Here's a purely hypothetical question. There is a lot of talk going around that quantum computers are close to finding the way to slow or stop the aging process. So tell me, are the rumors true? Are you on the verge of finding the Fountain of Youth?"

You reply, "Well, we don't have all the details yet, but it's true: our laboratories have been able to use gene therapy, CRISPR, and quantum computers to fix the errors caused by aging. We know that aging is the buildup of errors in our genes and cells. And now, we are finding the method to correct these errors, and hence slow down and perhaps reverse the aging process."

"That leads me to the final question. If you could have another lifetime, what would you like to be? For example, as a journalist, I would love to spend another lifetime being a novelist. What about you?"

"Well," you reply, "living several lifetimes is not such an outrageous possibility anymore. But if I could have another lifetime, I would like to apply quantum computers to solve the ultimate question about the universe. I mean, where did it come from? Why did we have the Big Bang? What happened before it? We humans are too primitive to solve these fantastic questions, but I bet that one day quantum computers may find the answer."

"Find the meaning of the universe? Wow, that's a tall order. But aren't you afraid of what the quantum computer may find?" she asks.

"Remember what happened at the end of *The Hitchhiker's Guide to the Galaxy*? After much anticipation and excitement, a giant supercomputer finally calculates the meaning of the universe. But the answer turns out to be the number forty-two. Well, that was a work of fiction. But now, I think we might be able to use quantum computers to crack this problem. Really," you reply.

After the interview, you shake her hand and thank Sarah for a great conversation. And then you discreetly ask her to join you for dinner. The article is a great success, informing millions of people how quantum computers have changed the economy, medicine, and our way of life. Another bonus was that you got to know Sarah better.

You are delighted to find out that you have much in common with her. You are both highly motivated and well informed. Later, you ask her to come visit the new Quantum Technologies video game parlor, where the most powerful quantum computers create the most realistic virtual games possible. You two have fun playing silly games, which create fantastic and exotic scenes via the powerful simulations of quantum computers. In one, you are exploring outer space. In another, a beach resort near the ocean. Next, the top of the highest mountain. You are amazed at how realistic they are, down to the smallest detail. But your favorite trip is to watch the full moon rising above the distant mountains. Watching the bright moon light up the forest, you can't help but feel close to nature.

You say to Sarah, "You know, watching the moon program, with astronauts beginning to explore the universe, that's how I first got interested in science."

Sarah replies, "Me too, but for me, the thrill was that one day I would see women walking on the moon."

Eventually, as you grow closer to her over time, you finally pop the question and ask her to marry you, and you are delighted when she says yes.

But where do you go for a honeymoon?

With all the news about the falling cost of space travel and consumers flying into outer space, she asks permission from her magazine for another story.

"I know just the place to have a honeymoon," says Sarah.

"I want to have a honeymoon on the moon."

QUANTUM PUZZLES

The cosmologist Stephen Hawking once said that physicists are the only scientists who can say the word "God" and not blush.

However, if you really wanted to see physicists blush, you could ask them deep philosophical questions for which there are no definitive answers.

Here is a short list of questions that will stump most physicists because they lie at the border between philosophy and physics. All of them affect the existence of quantum computers, and we will consider each of them in turn.

1. Did God have a choice in making the universe?

Einstein considered this to be one of the most profound and revealing questions one can possibly ask. Could God have created the universe in any other way?

2. Is the universe a simulation?

Are we just automatons living in a video game? Is everything we see and do a by-product of a computer simulation?

3. Do quantum computers compute in parallel universes?

Can we resolve the measurement problem for quantum computers by introducing a multiverse of universes?

4. Is the universe a quantum computer?

Can everything we see around us, from subatomic particles to galactic clusters, be evidence that the universe itself is a quantum computer?

Did God Have a Choice?

Einstein spent much of his life asking himself if the laws of the universe were unique, or were they just one of several possibilities. When we learn about quantum computers for the first time, their inner workings seem crazy and bizarre. It seems incredible that, at a fundamental level, electrons can exhibit such unrecognizable behavior, such as being in two places at the same time, tunneling through solid barriers, transmitting information faster than light, and instantly analyzing an infinite number of paths between any two points. You ask yourself, does the universe have to be this strange? If we had a choice, couldn't we rearrange the laws of physics to be more logical and sensible?

When Einstein was stuck on a problem, he would often say, "God is subtle, but not malicious." But when he had to face the paradoxes of quantum mechanics, sometimes Einstein would think, "Maybe God is malicious after all."

Throughout history, physicists have contemplated imaginary universes obeying a different set of fundamental laws, to see if the laws of nature are unique, and to see if it's possible to create a better universe from scratch.

Even philosophers have grappled with this cosmic question. Alfonso the Wise once said, "Had I been present at the creation, I would have given some useful hints for the better ordering of the universe."

Scottish judge and critic Lord Jeffrey would complain about all the imperfections in our universe. He would say, "Damn the solar system. Bad light, planets too distant, pestered with comets; feeble contrivance; could make a better [universe] myself."

However, scientists, no matter how hard they have tried, have been unable to improve on the laws of quantum physics. Usually, physicists find that alternatives to quantum mechanics yield universes that are unstable or suffer from some hidden fatal flaw.

To answer this philosophical question that fascinated Einstein, physicists often begin by listing the qualities that we want a universe to have.

First, and foremost, we want our universe to be stable. We don't want it to fall apart in our hands, leaving us with nothing.

Surprisingly, this criterion is extremely difficult to achieve. The simplest starting point might be to assume that we live in a common-sense, Newtonian world. This is the world that we are familiar with. Assume that this world is made of tiny atoms that are like miniature solar systems, with electrons going around a nucleus, obeying Newton's laws. This solar system would be stable if the electrons moved in perfect circles.

But if you slightly disturb one of these electrons, it might start to wobble and assume imperfect trajectories. This means that eventually these electrons will bump into each other or fall into the nucleus. Very quickly, the atom collapses and electrons fly everywhere. In other words, a Newtonian model of the atom is inherently unstable.

Consider what would happen with molecules. In a world governed only by classical mechanics, an orbit that goes around two nuclei is highly unstable and will rapidly fall apart as soon as you disturb it. Thus, molecules cannot exist in a Newtonian world, so there would be no complex chemicals. This universe, without stable atoms and molecules, eventually becomes a formless mist of random subatomic particles.

The quantum theory solves this problem, however, because the electron is described by a wave, and only discrete resonances of that wave can oscillate around the nucleus. Waves in which these electrons collide and fly apart are not allowed by the Schrödinger equation, so the atom is stable. In a quantum world, molecules too are stable because they are formed when the electron waves are shared between two different atoms, and a stable resonance forms that binds two atoms together. This provides the glue that can hold the molecule together.

So, in some sense, there is a "purpose" or "reason" for quantum mechanics and its strange features. Why is the quantum world so bizarre? Apparently to make matter stable and solid. Otherwise, our universe disintegrates.

This, in turn, has an important consequence for quantum computers. If one tries to modify the Schrödinger equation, which is the basis of the quantum computer, we expect that the modified quantum computer will yield nonsensical results, such as unstable matter. So in other words, the only way for quantum computers to create stable universes is to begin with the Schrödinger equation. A quantum computer is unique. There may be many ways in which matter can be assembled to create a quantum computer (e.g., with different types of atoms) but there is only one way in which the quantum computer can perform its calculations and still describe stable matter.

So if we want a quantum computer that manipulates electrons, light, and atoms, then we are probably left with a unique architecture for a quantum computer.

Universe as a Simulation

Anyone who has seen the movie *The Matrix* knows that Neo is the Chosen One. He has superpowers. He can soar into the heavens. He can dodge speeding bullets or make them stop midflight. He can instantly learn karate with the push of a button. And he can walk through mirrors.

All this is possible because Neo actually lives in a fictitious, computer-generated simulation. Like living in a video game, "reality" is actually an imaginary world.

But this raises the question: With computer power growing exponentially, is it possible that our world is actually a simulation, and that the "reality" we know is a video game played by someone else? Are we just lines of code, until someone finally hits the delete button and ends the charade? And if a classical computer is not powerful enough to simulate reality, then can a quantum computer do this?

Let's first ask a simpler question: Can a classical universe like the one described above be a Newtonian simulation?

Consider for a moment an empty glass bottle. The air inside that bottle may contain more than 10^{23} atoms. To model this exactly with a classical computer, you would need to manipulate 10^{23} bits of information, which is far beyond what a classical computer can do. To create a perfect simulation of the atoms in that bottle, you would also have to know the position and velocity of all these atoms. Now imagine trying to simulate the weather on the earth. You have to know the humidity, air pressure, temperature, and wind velocity of the air around the planet. Very quickly, you exhaust the memory capacity of any known classical computer.

In other words, the smallest object that can simulate the weather is the weather itself.

Another way to see this problem is to consider what is called the butterfly effect. If a butterfly flaps its wings, it might create a wave of air that might, if the conditions are just right, eventually cascade into a powerful wind. This in turn might eventually hit the tipping point of a cloud, causing a rainstorm. This is a result from chaos theory, which says that although air molecules may obey Newtonian laws, the combined effect of trillions of air molecules are chaotic and unpredictable. So predicting the precise probability that a rainstorm can form is nearly impossible. Although the path of a single molecule can be determined, the collective motion of trillions of air molecules is beyond the reach of any digital computer. A simulation is again impossible.

But what about quantum computers?

The situation gets much worse if we try to model the weather with a quantum computer. If we have a quantum computer with 300 qubits, then we have 2^{300} states in the quantum computer, which is larger than the states of the universe. Surely a quantum computer has enough memory to encode all of "reality" as we know it.

Not necessarily. Think of a complex protein molecule, which may have thousands of atoms. For a quantum computer to simulate just one protein molecule without any approximation whatsoever, we have to have many more states than there are in the universe. But our body may have billions of these protein molecules. So to truly simulate all the protein molecules found in our body, we need in principle to have billions of quantum computers. Once again, the smallest object that can simulate the universe is the universe itself. It is simply impractical to assemble billions upon billions of quantum computers to simulate a complex quantum phenomenon.

The only "reality" that might actually be simulated is one that is not perfect, but has many gaps and imperfections. This could reduce the number of states that have to be simulated. If the simulation is not perfect, then it might actually exist. For example, the simulation may have areas that are incomplete. The "sky" you see above you may have rips and tears, like an old movie set. Or if you are a deep-sea diver, you might think your world is the entire ocean, until you bump into a glass wall, and then you realize that your world is just a small simulation of the ocean. So a universe with imperfections such as these is certainly possible.

Parallel Universes

It used to be that Hollywood and the comics could create exciting, imaginary universes by taking their characters into outer space. But since we have been sending rockets into outer space for over fifty years, this is a bit passé. So science fiction writers need a new, cutting-edge playground for their fantastic plots, and now it is the multiverse. Many recent blockbusters are set in parallel universes, where the superhero or villain exists in multiple realities.

In the past, whenever I would see a science fiction movie, I used to count how many laws of physics were violated. I stopped doing that when I remembered Arthur C. Clarke's words: "Any sufficiently advanced technology is indistinguishable from magic." So if a movie apparently violated some known law of physics, perhaps this law of physics will one day be shown to be incorrect or incomplete.

But now, with the movies entering the multiverse of parallel universes, I have to think again to see if any physical laws are violated. In this case, the movies are actually following the lead of theoretical physicists, who are taking the idea of the multiverse seriously.

The reason for this is that Hugh Everett's many worlds theory is making a comeback. As mentioned earlier, Everett's many worlds theory is perhaps the simplest and most elegant way to resolve the measurement problem. By simply dropping the last postulate of quantum mechanics, that the wave function describing quantum behavior collapses upon observation, the many worlds theory is the quickest way to resolve the paradox it poses.

But there is a price to pay for allowing the electron wave to proliferate. If the Schrödinger wave is allowed to freely move by itself, without collapsing, then it will divide an infinite number of times, creating an infinite cascade of possible universes. So instead of collapsing down to one universe, we let an infinite number of parallel universes constantly split apart.

There is no universal consensus among physicists about these parallel universes. For example, David Deutsch believes this is the essential reason why quantum computers are so powerful, because they calculate simultaneously in different parallel universes. This takes us back to Schrödinger's old paradox, where a cat in a box can be simultaneously dead and alive.

Stephen Hawking, when asked about this frustrating problem, would say, "Whenever I hear Schrödinger's cat, I reach for my gun."

But there is an alternative theory that is also being considered, called decoherence theory, that states that interactions with the external environment cause the wave to collapse, i.e., the wave collapses all by itself once it touches the environment, because the environment has already decohered.

For example, this means that Schrödinger's paradox can be simply resolved. The original problem was that before you opened the box, you could not tell if the cat is dead or alive. The traditional answer is that the cat is neither dead nor alive until you open the box. This new theory says the cat's atoms are already in contact with the random atoms floating in the box, so the cat has already decohered even before you open the box. So the cat is already dead or alive (but not both).

In other words, according to the traditional Copenhagen interpretation, the cat decoheres only when you open the box and you make a measurement. In the decoherence approach, however, the cat has already decohered, because the air molecules have touched the cat's wave, causing it to collapse. The cause of the wave's collapse in the decoherence approach replaces the experimenter who opens the box with the air in the box.

Usually debates in physics are resolved by doing an experiment. Physics is not ultimately based on speculation and conjecture. The decisive factor is hard evidence. But I imagine that decades from now, physicists will still debate this question, because there is no decisive experiment that can rule out one of these interpretations, at least not yet.

However, I personally think there is a flaw in the decoherence approach. This approach has to make a distinction between the environment, i.e., the air (which is decoherent), and the object being studied (the cat). In the Copenhagen approach, decoherence is introduced by the experimenter. In the decoherence approach, it is introduced by interactions with the environment.

However, once we introduce a quantum theory of gravity, the smallest unit that we quantize is the universe itself. There is no distinction between the experimenter, the environment, and the cat. They are all part of one gigantic wave function, the wave function of the universe, which cannot be separated into various pieces.

In this quantum gravity approach, there is no real distinction between waves that are coherent and waves in the air that are decoherent. The difference is only one of degree. (For example, at the Big Bang, the entire universe was coherent before the explosion. So even today, 13.8 billion years later, there is a bit of coherence that we can still find between the cat and the air.)

So this approach banishes decoherence and reverts back to the Everett interpretation. Unfortunately, there is no experiment that can tell the difference between these various approaches. Both approaches give the same quantum mechanical result. They differ in the interpretation of the result, which is philosophical.

This means that whether we use the Copenhagen interpretation, the decoherence approach, or the many worlds theory, we get the same experimental results, so all three approaches are experimentally equivalent.

One possible difference between these three approaches is that, in the many worlds interpretation, it might be possible to move between different parallel universes. But if one does the calculation, the probability of being able to do so is so small, we cannot experimentally verify it. Usually, we have to wait longer than the lifetime of the universe to enter another parallel universe.

Is the Universe a Quantum Computer?

Now let us analyze whether the universe itself is a quantum computer.

We recall that Babbage asked himself a well-defined question: How powerful can you make an analog computer? What are the limits to what you can compute with mechanical gears and levers?

Turing extended this question by asking himself another: How powerful can you make a digital computer? What are the limits to computation with electronic components?

Therefore, it is natural to ask next: How powerful can you make a quantum computer? What are the limits to computation if we can manipulate individual atoms? And since the universe is made of atoms: Is the universe itself a quantum computer?

The physicist who proposed this idea is Seth Lloyd of MIT. He is one of a handful of physicists who were there at the very beginning when quantum computers were first being created.

I asked Lloyd how he became involved with quantum computers. He told me that when he was a youth, he was fascinated by numbers. He was especially interested in the fact that, with just a few numbers, one could describe a vast quantity of objects in the real world using the rules of mathematics.

When he went to graduate school, however, he was faced with a problem. On one hand, there were bright physics students doing string theory and elementary particle physics. On the other hand, there were students doing computer science. He was caught in between, because he wanted to work on quantum information, which was midway between particle physics and computer science.

In elementary particle physics, the ultimate unit of matter is the particle, such as the electron. In information theory, the ultimate unit of information is the bit. So he has been concerned about the relationship between particles and bits, which leads us to quantum bits.

His controversial idea is that the universe is a quantum computer. At first, that may sound outlandish. When we think of the universe, we think of stars, galaxies, planets, animals, people, DNA. But when we think of a quantum computer, we think of a machine. How can they be the same?

Actually, there is a profound relationship between the two. It is possible to create a Turing machine that can contain all the Newtonian laws of the universe.

Think, for example, of a toy train sitting on a miniature train track. The track is divided into a long sequence of squares, in which we can place the number 0 or 1. 0 means that there is no train in that track, and 1 means the toy train is in that track. Now let us move the train, square by square. Each time we move the train by one square, we replace a 0 with a 1. In this way, the train can move smoothly along the track. The number 1 locates the position of the toy train.

Now, let us replace the railroad track with a digital tape, with 0s and 1s. Replace the toy train with the processor. Each time the processor moves by one square, we replace 0 with a 1.

In this way, we can take a toy train and convert it into a Turing machine. In other words, a Turing machine can simulate Newton's laws of motion, which is the foundation of classical physics.

We can also modify the toy train to describe accelerations and more complex motions. Each time we move the toy train, we can increase the separation between the 1s, so the train is speeding up. We can also generalize the toy train riding along a 3D track, or lattice. In this way, we can encode all the laws of Newtonian mechanics.

So now we can precisely make the link between a Turing machine and Newton's laws. A classical universe can be encoded by a Turing machine.

Next, we can generalize this to quantum computers. Instead of a toy train, which contains 0s and 1s, we replace it with a toy train carrying a compass. Its needle can point north, where we call it 1, or south, where we call it a 0, or any angle between the two, where it represents the superposition of north and south. So as the toy train moves down the track, the needle moves in different directions, according to the Schrödinger equation.

(If one wants to include entanglement, then one adds several compasses on the toy train. All these compass needles can move in different ways as the train moves down the track, according to the rules of the processor.)

As the toy train moves, the needle of the compass starts to spin. The motion of the needle traces out the information contained in the Schrödinger wave equation. So in this way, we can derive the wave equation using this toy train.

The point here is that a quantum Turing machine can encode the laws of quantum mechanics, which in turn rule the universe. In this sense, a quantum computer can encode the universe. So the relationship between a quantum computer and the universe is that the former can codify the latter. Thus, strictly speaking, the universe is not a quantum computer, but all phenomena within the universe can be codified by a quantum computer.

But since all interactions at the microscopic level are governed by quantum mechanics, it means that quantum computers can simulate any phenomenon of the physical world, from subatomic particles, DNA, and black holes to the Big Bang.

The playground for quantum computers is the universe itself. So if we can truly understand a quantum Turing machine, then perhaps we can truly understand the universe as well.

Only time can tell.

ACKNOWLEDGMENTS

I would like to thank first and foremost my literary agent, Stuart Krichevsky, who has been with me all these long years, helping shepherd my books from the drawing board to the marketplace. I trust his unerring judgment in all literary matters. His sound advice has helped make my books a success.

I would also like to thank my editor, Edward Kastenmeier. He has always provided wise judgment in all editorial matters. At every step of the way, he has helped sharpen the focus of the book and make it more readily accessible.

I would also like to thank the many Nobel laureates whom I have consulted with or interviewed who provided invaluable advice:

Richard Feynman
Steven Weinberg
Yoichiro Nambu
Walter Gilbert
Henry Kendall
Leon Lederman
Murray Gell-Mann
David Gross
Frank Wilczek

Joseph Rotblat
Henry Pollack
Peter Doherty
Eric Chivian
Gerald Edelman
Anton Zeilinger
Svante Pääbo
Roger Penrose

I would also like to thank these prominent scientists, who have been leaders in scientific research or directors of major scientific laboratories, and who have generously shared their wisdom with me:

Marvin Minsky
Francis Collins
Rodney Brooks
Anthony Atala
Leonard Hayflick
Carl Zimmer
Stephen Hawking
Edward Witten
Michael Lemonick
Michael Shermer
Seth Shostak
Ken Croswell
Brian Greene
Neil deGrasse Tyson
Lisa Randall
Leonard Susskind

Lastly, I would like to thank the more than 400 scientists whom I have interviewed over the years, whose insights were invaluable in writing this book.

NOTES

Chapter 1: End of the Age of Silicon

3   "the threshold of a new era of machines": Gordon Lichfield, "Inside the Race to Build the Best Quantum Computer on Earth," *MIT Technology Review,* February 26, 2020, 1–23.

3   "I think it's going to be": Yuval Boger, interview with Dr. Robert Sutor, *The Qubit Guy's Podcast,* October 27, 2021; www.classiq.io/insights/podcast-with-dr-robert-sutor.

5   "It's no longer a matter of if": Matt Swayne, "Zapata Chief Says Quantum Machine Learning Is a When, Not an If," *The Quantum Insider,* July 16, 2020; www.thequantuminsider.com/2020/07/16/zapata-chief-says-quantum-machine-learning-is-a-when-not-an-if/.

6   "Companies in the industries": Daphne Leprince-Ringuet, "Quantum Computers Are Coming, Get Ready for Them to Change Everything," *ZD Net,* November 2, 2020; www.zdnet.com/article/quantum-computers-are-coming-get-ready-for-them-to-change-everything/.

7   "We are excited to investigate": Dashveenjit Kaur, "BMW Embraces Quantum Computing to Enhance Supply Chain," *Techwire/Asia,* February 1, 2021; www.techwireasia.com/2021/02/bmw-embraces-quantum-computing-to-enhance-supply-chain/.

7   "It is not that machines": Cade Metz, "Making New Drugs with a Dose of Artificial Intelligence," *The New York Times,* February 5, 2019; www.nytimes.com/2019/02/05/technology/artificial-intelligence-drug-research-deepmind.html.

8   "That's a pretty terrifying": Ali El Kaafarani, "Four Ways That Quantum Computers Can Change the World," *Forbes,* July 30, 2021; www.forbes.com/sites/forbestechcouncil/2021/07/30/four-ways-quantum-computing-could-change-the-world/?sh=7054e3664602.

9 "Those running blockchain": "How Quantum Computers Will Transform These 9 Industries," *CB Insights,* February 23, 2021; www.cbinsights.com /research/quantum-computing-industries-disrupted/.

11 "We've squeezed": Matthew Hutson, "The Future of Computing," *ScienceNews;* www.sciencenews.org/century/computer-ai-algorithm-moore-law-ethics.

11 "It looks like nothing": James Dargan, "Neven's Law: Paradigm Shift in Quantum Computers," *Hackernoon,* July 1, 2019; www.hackernoon.com/nevens -law-paradigm-shift-in-quantum-computers-e6c429ccd1fc.

18 Jeremy O'Brien: Nicole Hemsoth, "With $3.1 Billion Valuation, What's Ahead for PsiQuantum?," *The Next Platform,* July 27, 2021; www.nextplatform.com /2021/07/27/with-3-1b-valuation-whats-ahead-for-psiquantum/.

Chapter 2: End of the Digital Age

27 "Ada saw something": "Our Founding Figures: Ada Lovelace," *Tetra Defense,* April 17, 2020; www.tetradefense.com/cyber-risk-management/our-founding -figures-ada-lovelace/.

28 "the engine might compose": "Ada Lovelace," Computer History Museum; www.computerhistory.org/babbage/adalovelace/.

30 "has clever boys": Colin Drury, "Alan Turing: The Father of Modern Computing Credited with Saving Millions of Lives," *The Independent,* July 15, 2019; www.independent.co.uk/news/uk/home-news/alan-turing-ps50-note -computers-maths-enigma-codebreaker-ai-test-a9005266.html.

36 "I believe that in about fifty": Alan Turing, "Computing Machinery and Intelligence," *Mind* 59 (1950): 433–60; https://courses.edx.org/asset -v1:MITx+24.09x+3T2015+type@asset+block/5_turing_computing _machinery_and_intelligence.pdf.

Chapter 3: Rise of the Quantum

41 "A new scientific truth": Peter Coy, "Science Advances One Funeral at a Time, the Latest Nobel Proves It," *Bloomberg,* October 10, 2017; www.bloomberg.com /news/articles/2017-10-10/science-advances-one-funeral-at-a-time-the-latest -nobel-proves-it.

45 "The fundamental laws": BrainyQuote; https://www.brainyquote.com/quotes /paul_dirac_279318.

53 "I will never forget": Jim Martorano, "The Greatest Heavyweight Fight of All Time," *TAP into Yorktown,* August 24, 2022; https://www.tapinto.net/towns /yorktown/articles/the-greatest-heavyweight-fight-of-all-time.

54 "It was the greatest debate": quoted in Denis Brian, *Einstein* (New York: Wiley, 1996), 516.

Chapter 4: Dawn of Quantum Computers

75   Unfortunately, Everett's idea: See Michio Kaku, *Parallel Worlds: The Science of Alternative Universes and Our Future in the Cosmos* (New York: Anchor, 2006).

76   "undescribably [sic] stupid": Stefano Osnaghi, Fabio Freitas, Olival Freire Jr., "The Origin of the Everettian Heresy," *Studies in History and Philosophy of Modern Physics* 40, no. 2 (2009): 17.

Chapter 5: The Race Is On

88   "We believe that we": Stephen Nellis, "IBM Says Quantum Chip Could Beat Standard Chips in Two Years," Reuters, November 15, 2021; www.reuters.com /article/ibm-quantum-idCAKBN2I00C6.

92   "My first impression was, Wow!": Emily Conover, "The New Light-Based Quantum Computer Jiuzhang Has Achieved Quantum Supremacy," *Science News,* December 3, 2020; https://www.sciencenews.org/article/new-light-based -quantum-computer-jiuzhang-supremacy.

94   "For a long time, photonics": "Xanadu Makes Photonic Quantum Chip Available Over Cloud Using Strawberry Fields & Pennylane Open-Source Tools Available on Github," *Inside Quantum Technology News,* March 8, 2021; www.insidequantumtechnology.com/news-archive/xanada-makes-photonic -quantum-chip-available-over-cloud-using-strawberry-fields-pennylane-open -source-tools-available-on-github/.

Chapter 6: The Origin of Life

105   "From the moment": Walter Moore, *Schrödinger: Life and Thought* (Cambridge University Press, 1989), 403.

112   "As molecules get larger": Leah Crane, "Google Has Performed the Biggest Quantum Chemistry Simulation Ever," *New Scientist,* December 12, 2019; www.newscientist.com/article/2227244-google-has-performed-the-biggest -quantum-chemistry-simulation-ever/.

112   "predicting the behavior": Jeannette M. Garcia, "How Quantum Computing Could Remake Chemistry," *Scientific American,* March 15, 2021; https://www .scientificamerican.com/article/how-quantum-computing-could-remake -chemistry/.

113   "The atoms are quantum": Crane.

114   "That's a result": Ibid.

Chapter 7: Greening the World

117   "I really want to know": Alan S. Brown, "Unraveling the Quantum Mysteries of Photosynthesis," The Kavli Foundation, December 15, 2020; www.kavlifoundation.org/news/unraveling-the-quantum-mysteries-of -photosynthesis.

121   "The excitation effectively": Peter Byrne, "In Pursuit of Quantum Biology with Birgitta Whaley," *Quanta Magazine,* July 30, 2013; www.quantamagazine.org/in -pursuit-of-quantum-biology-with-birgitta-whaley-20130730/.

123   "Our goal is to close": Katherine Bourzac, "Will the Artificial Leaf Sprout to Combat Climate Change?," *Chemical & Engineering News,* November 21, 2016; https://cen.acs.org/articles/94/i46/artificial-leaf-sprout-combat-climate.html.

123   "quantum computers may be able to accelerate": Ali El Kaafarani, "Four Ways Quantum Computing Could Change the World," *Forbes,* July 30, 2021; www.forbes.com/sites/forbestechcouncil/2021/07/30/four-ways-quantum -computing-could-change-the-world/?sh=398352d14602.

123   "We did a complete": Katharine Sanderson, "Artificial Leaves: Bionic Photosynthesis as Good as the Real Thing, *New Scientist,* March 2, 2022; www.newscientist.com/article/mg25333762-600-artificial-leaves-bionic -photosynthesis-as-good-as-the-real-thing/.

Chapter 8: Feeding the Planet

132   "Using today's supercomputers": "What Is Quantum Computing? Definition, Industry Trends, & Benefits Explained," *CB Insights,* January 7, 2021; https:// www.cbinsights.com/research/report/quantum-computing/?utm_source= CB+Insights+Newsletter&utm_campaign=0df1cb4286-newsletter_general _Sat_20191115&utm_medium=email&utm_term=0_9dc0513989-0df1cb4286 -88679829.

134   "I think we're at an inflection": Allison Lin, "Microsoft Doubles Down on Quantum Computing Bet," Microsoft, *The AI Blog,* November 20, 2016; https:// blogs.microsoft.com/ai/microsoft-doubles-quantum-computing-bet/.

135   In fact, Google's CEO: Stephen Gossett, "10 Quantum Computing Applications and Examples," *Built In,* March 25, 2020; https://builtin.com/hardware /quantum-computing-applications.

Chapter 9: Energizing the World

145   "It's a very research-oriented": Holger Mohn, "What's Behind Quantum Computing and Why Daimler Is Researching It," Mercedes-Benz Group,

August 20, 2020; https://group.mercedes-benz.com/company/magazine
/technology-innovation/quantum-computing.html.

145   "It could become the best": Ibid.

Chapter 11: Gene Editing and Curing Cancer

164   "In recent years": Liz Kwo and Jenna Aronson, "The Promise of Liquid
Biopsies for Cancer Diagnosis," *American Journal of Managed Care,* October 11,
2021; www.ajmc.com/view/the-promise-of-liquid-biopsies-for-cancer
-diagnosis.

174   "In theory": Clara Rodríguez Fernández, "Eight Diseases CRISPR Technology
Could Cure," *Labiotech,* October 18, 2021; https://www.labiotech.eu/best
-biotech/crispr-technology-cure-disease/.

177   "The hope is that": Viviane Callier, "A Zombie Gene Protects Elephants from
Cancer," *Quanta Magazine,* November 7, 2017; www.quantamagazine.org/a
-zombie-gene-protects-elephants-from-cancer-20171107/.

Chapter 12: AI and Quantum Computers

179   "An attempt will": Gil Press, "Artificial Intelligence (AI) Defined," *Forbes,*
August 27, 2017; https://www.forbes.com/sites/gilpress/2017/08/27/artificial
-intelligence-ai-defined/.

182   "I think AI": Stephen Gossett, "10 Quantum Computing Applications and
Examples," *Built In,* March 25, 2020; https://builtin.com/hardware/quantum
-computing-applications.

191   "We have been stuck": "AlphaFold: A Solution to a 50-Year-Old Grand
Challenge in Biology," DeepMind, November 30, 2020; www.deepmind.com
/blog/alphafold-a-solution-to-a-50-year-old-grand-challenge-in-biology.

191   "We were able": Cade Metz, "London A.I. Lab Claims Breakthrough That
Could Accelerate Drug Discovery," *The New York Times,* November 30, 2020;
https://www.nytimes.com/2020/11/30/technology/deepmind-ai-protein
-folding.html.

196   "I believe this shows": Ron Leuty, "Controversial Alzheimer's Disease Theory
Could Pinpoint New Drug Targets," *San Francisco Business Times,* May 6,
2019; www.bizjournals.com/sanfrancisco/news/2019/05/01/alzheimers-disease
-prions-amyloid-ucsf-prusiner.html.

196   "The measurement of misfolded": German Cancer Research Center, "Protein
Misfolding as a Risk Marker for Alzheimer's Disease," *ScienceDaily,* October 15,
2019; www.sciencedaily.com/releases/2019/10/191015140243.htm.

196   "Everyone is now": "Protein Misfolding as a Risk Marker for Alzheimer's
Disease—Up to 14 Years Before the Diagnosis," Bionity.com, October 17, 2019;

www.bionity.com/en/news/1163273/protein-misfolding-as-a-risk-marker-for
-alzheimers-disease-up-to-14-years-before-the-diagnosis.html.

Chapter 13: Immortality

209   "As we get older": Mallory Locklear, "Calorie Restriction Trial Reveals Key
Factors in Enhancing Human Health," *Yale News*, February 10, 2022; www
.news.yale.edu/2022/02/10/calorie-restriction-trial-reveals-key-factors
-enhancing-human-health.

211   "If diseases happen": Kashmira Gander, " 'Longevity Gene' That Helps Repair
DNA and Extend Life Span Could One Day Prevent Age-Related Diseases in
Humans," *Newsweek,* April 23, 2019; www.newsweek.com/longevity-gene-helps
-repair-dna-and-extend-lifespan-could-one-day-prevent-age-1403257.

212   "If you see something": Antonio Regalado, "Meet Altos Labs, Silicon Valley's
Latest Wild Bet on Living Forever," *MIT Technology Review,* September 4, 2021;
www.technologyreview.com/2021/09/04/1034364/altos-labs-silicon-valleys-jeff
-bezos-milner-bet-living-forever/.

213   "There are hundreds": Ibid.

213   "You can take a cell": Antonio Regalado, "Meet Altos Labs, Silicon Valley's
Latest Wild Bet on Living Forever," *MIT Technology Review,* September 4, 2021;
www.technologyreview.com/2021/09/04/1034364/altos-labs-silicon-valleys-jeff
-bezos-milner-bet-living-forever/.

214   "I remember the day": Allana Akhtar, "Scientists Rejuvenated the Skin of a 53
Year Old Woman to That of a 23 Year Old's in a Groundbreaking Experiment,"
Yahoo News, April 8, 2022; www.yahoo.com/news/scientists-rejuvenated-skin
-53-old-175044826.html.

Chapter 14: Global Warming

227   "Quantum computers also hold": Ali El Kaafarani, "Four Ways Quantum
Computing Could Change the World," *Forbes,* July 30, 2021; www.forbes.com
/sites/forbestechcouncil/2021/07/30/four-ways-quantum-computing-could
-change-the-world/?sh=398352d14602.

229   "Sea-level rise": Doyle Rice, "Rising Waters: Climate Change Could Push a
Century's Worth of Sea Rise in US by 2050, Report Says," *USA Today,* February
15, 2022; https://www.usatoday.com/story/news/nation/2022/02/15/us-sea-rise
-climate-change-noaa-report/6797438001/.

230   "This report supports": "U.S. Coastline to See up to a Foot of Sea Level Rise by
2050," National Oceanic and Atmospheric Administration, February 15, 2022;
https://www.noaa.gov/news-release/us-coastline-to-see-up-to-foot-of-sea
-level-rise-by-2050.

232   "The eastern ice shelf": David Knowles, "Antarctica's 'Doomsday Glacier'

Is Facing Threat of Imminent Collapse, Scientists Warn," *Yahoo News,* December 14, 2021; https://news.yahoo.com/antarcticas-doomsday-glacier-is -facing-threat-of-imminent-collapse-scientists-warn-220236266.html.

236   For the Fourth Assessment Report: Intergovernmental Panel on Climate Change, *Climate Change 2007 Synthesis Report: A Report of the Intergovernmental Panel on Climate Change;* www.ipcc.ch.

Chapter 15: The Sun in a Bottle

246   "Fusion is not": Jonathan Amos, "Major Breakthrough on Nuclear Fusion Energy," *BBC News,* September 9, 2022; www.bbc.com/news/science -environment-60312633.

247   "The JET experiments put us": Claude Forthomme, "Nuclear Fusion: How the Power of Stars May Be Within Our Reach," *Impakter,* February 10, 2022; www .impakter.com/nuclear-fusion-power-stars-reach/.

247   "It's a landmark": Jonathan Amos, "Major Breakthrough on Nuclear Fusion Energy," *BBC News,* September 9, 2022; www.bbc.com/news/science -environment-60312633.

248   "This magnet will": "Multiple Breakthroughs Raise New Hopes for Fusion Energy," Global BSG, January 27, 2022; www.globalbsg.com/multiple -breakthroughs-raise-new-hopes-for-fusion-energy/.

248   "It's a big deal": Catherine Clifford, "Fusion Gets Closer with Successful Test of a New Kind of Magnet at MIT Start-up Backed by Bill Gates," CNBC, September 8, 2021; www.cnbc.com/2021/09/08/fusion-gets-closer-with -successful-test-of-new-kind-of-magnet.html.

253   "I think AI will play": "Nuclear Fusion Is One Step Closer with New AI Breakthrough," *Nation World News,* September 13, 2022; www .nationworldnews.com/nuclear-fusion-is-one-step-closer-with-new-ai -breakthrough/.

Chapter 16: Simulating the Universe

265   "A scene of almost unspeakable": "The World Should Think Better About Catastrophic and Existential Risks," *The Economist,* June 25, 2020; www .economist.com/briefing/2020/06/25/the-world-should-think-better-about -catastrophic-and-existential-risks.

276   So far, the leading: For a discussion of string theory, see Michio Kaku, *The God Equation: The Quest for a Theory of Everything* (New York: Anchor, 2022).

SELECTED READING

For those who have some familiarity with computer programming, the following texts may prove useful:

Bernhardt, Chris. *Quantum Computing for Everyone.* Cambridge: MIT Press, 2020.

Edwards, Simon. *Quantum Computing for Beginners.* Monee, IL, 2021.

Grumbling, Emily, and Mark Horowitz, eds. *Quantum Computing: Progress and Prospects.* Washington, DC: National Academy Press, 2019.

Jaeger, Lars. *The Second Quantum Revolution.* Switzerland: Springer, 2018.

Mermin, N. David. *Quantum Computer Science: An Introduction.* Cambridge: Cambridge University Press, 2016.

Rohde, Peter P. *The Quantum Internet: The Second Quantum Revolution.* Cambridge: Cambridge University Press, 2021.

Sutor, Robert S. *Dancing with Qubits: How Quantum Computing Works and How It Can Change the World.* Birmingham, UK: Packt, 2019.

ILLUSTRATION CREDITS

Page 24: Freeth, T., Higgon, D., Dacanalis, A., et al. A Model of the Cosmos in the Ancient Greek Antikythera Mechanism. *Sci Rep* 11, 5821 (2021).

Page 32: Mapping Specialists Ltd.

Page 43: Mapping Specialists Ltd.

Page 50: Mapping Specialists Ltd.

Page 55: Mapping Specialists Ltd.

Page 68: Mapping Specialists Ltd.

Page 81: Mapping Specialists Ltd.

Page 89: Andrew Lindemann, courtesy IBM

Page 91: Mapping Specialists Ltd.

Page 188: Mapping Specialists Ltd.

Page 243: Mapping Specialists Ltd.

Page 270: Mapping Specialists Ltd.

INDEX

THE FUTURE OF HUMANITY
Our Destiny in the Universe

We are entering a new Golden Age of space exploration. With irrepressible enthusiasm and a deep understanding of the cutting-edge research in space travel, world-renowned physicist and futurist Dr. Michio Kaku reveals the developments in robotics, nanotechnology, and biotechnology that may allow us to terraform and build habitable cities on Mars and beyond. He then journeys out of our solar system and discusses how new technologies such as nanoships, laser sails, and fusion rockets may actually make interstellar travel a possibility. We travel beyond our galaxy, and even beyond our universe, as Kaku investigates some of the hottest topics in science today, including warp drive, wormholes, hyperspace, parallel universes, and the multiverse. Ultimately, he shows us how humans may someday achieve a form of immortality and be able to leave our bodies entirely, laser porting to new havens in space.

Physics

ALSO AVAILABLE

Beyond Einstein
Hyperspace
Parallel Worlds
Physics of the Impossible
Visions

VINTAGE BOOKS
Available wherever books are sold.
vintagebooks.com